Synthesis Lectures on Digital Circuits & Systems

Series Editor

Mitchell A. Thornton, Southern Methodist University, Dallas, USA

This series includes titles of interest to students, professionals, and researchers in the area of design and analysis of digital circuits and systems. Each Lecture is self-contained and focuses on the background information required to understand the subject matter and practical case studies that illustrate applications. The format of a Lecture is structured such that each will be devoted to a specific topic in digital circuits and systems rather than a larger overview of several topics such as that found in a comprehensive handbook. The Lectures cover both well-established areas as well as newly developed or emerging material in digital circuits and systems design and analysis.

Matteo Petrini

Mixed-Signal Generic Testing in Photonic Integration

 Springer

Matteo Petrini
Cisco Photonics S.r.l.
Vimercate, Italy

ISSN 1932-3166   ISSN 1932-3174 (electronic)
Synthesis Lectures on Digital Circuits & Systems
ISBN 978-3-031-60810-0   ISBN 978-3-031-60811-7 (eBook)
https://doi.org/10.1007/978-3-031-60811-7

This Springer imprint is published by the registered company Springer Nature Switzerland AG
The registered company address is: Gewerbestrasse 11, 6330 Cham, Switzerland

If disposing of this product, please recycle the paper.

Introduction

Photonics is expected to bring the same revolution that microelectronics have brought in the last century. Specifically, the spread of optical components is being boosted by their miniaturization and integration. Thus, in the recent decades, the locution Photonic Integrated Circuit (PIC) arose.

These PICs are currently exploited for several different applications, such as metrology, sensing (especially in the field of biology), environmental monitoring, quantum computing, and their development is occurring very quickly. However, traditionally, Optical Communications have always been the target market for this kind of technology. In the framework of telecommunications, in fact, the trend is to reduce the energy consumption per bit and keep the bandwidth fairly large, given the enormous amount of data that are currently transmitted. These features match the characteristics of photonic components, that are rapidly replacing electronic circuits.

To sustain this "electronics-to-photonics" transition, the integration density must constantly increase, following a trend that has been called "Moore's law for photonics" [1]. However, it is worth emphasizing that integration drivers are indeed different between microelectronics and photonics. While in microelectronics the number of building blocks (i.e., transistors) per unit area is doubling every 2 years (limited only by lithography advances), PIC sizes have remained almost constant in the last two decades and are mostly bounded by the choice of semiconductor/dielectric materials and by geometrical constraints, to achieve propagation with acceptable loss.

The exploitation of high refractive index platforms (such as silicon or indium phosphide) and the use of resonance-based building blocks definitely contribute to increasing the integration density of PICs.

Currently, we are observing that the implementation of several functionalities in a single chip is not enough. The reconfigurability and programmability of those functions upon the users' needs are becoming requirements of primary importance. Thus, there is a paradigm shift: a single device is turning into a "system-on-chip".

Clearly, the density-increasing and tuneable functions increase the complexity of the PIC itself and of all the processes aimed at producing it. Among these, we can certainly find the testing stage.

The main purpose of this book is the presentation of a collection of practices to perform reliable and cost-effective photonic testing.

In Chap. 2, we report an example of a dense and inherently complex reconfigurable PIC, specifically a four-channel silicon photonic Tuneable Optical Add/Drop Multiplexer (TOADM). Starting from the desired specifications, we discuss its design and outcomes of its characterization. Measurements have been performed at both the device and system levels.

This will be the Device Under Test (DUT) for the development of our techniques, that are effectively executed by a proper electronic system. In Chap. 3, in fact, we focus on the electronics required to perform electro-optical testing of the DUT. We couple a commercial controller with a pure analog (custom) board for signal conditioning. Then, by means of a Probe Card (PC), we feed the DUT with these generated testing signals and we perform an electrical characterization of the whole chip. Since we are dealing with thermo-optic tuners (embedded in the PIC, for calibration and reconfiguration purposes), one of the outcomes of this electrical characterization is the quantitative evaluation of thermal crosstalk.

This infrastructure also enables for the execution of "pre-calibration" algorithms, which are useful for tuning the DUT before performing optical validation. Testing an "out-of-fab" complex PIC is, in fact, meaningless, since due to fabrication imperfections, its time and/or frequency domain behavior may be severely distorted.

In this regard, in Chap. 4, we propose a novel paradigm for spectral testing/tuning of complex filters. Instead of relying on a priori determined spectral masks, and on wavelength sweeps/scans with tuneable lasers and spectrum analyszrs, we color a white noise source (by using a reference filter) and we feed the DUT on this and by measuring power at its output port(s) we are able to assess how far it is from the reference. This information is also useful as a feedback signal for closed-loop tuning.

Once the PIC is electrically validated and it is optically compliant with the specifications (after the pre-calibration stage), the working points of its actuators can be conveniently stored in a LookUp Table (LUT), which must be updated in case of environmental changes and/or different user needs.

Nevertheless, the considered DUT is a silicon photonic circuit and is, thus, by definition sensitive to high-power input signals. In Chap. 5, we analyze the impact of nonlinearities and how to counteract them, by using the available actuators. In so doing, the LUT could be advantageously updated even in this extreme case.

Finally, in Chap. 6, we extend these results and we shift from filters/multiplexers to optical delay lines. The very same surrounding electronics have been exploited and the very same pattern (in terms of optical pre-calibration) has been applied.

In the end, we foresee future directions for this specific research work and desirable trends in the field of testing, that would boost Photonics in the near future.

Publications

A list of the publications and original contributions of this thesis is here provided.

Journals

1. M. Milanizadeh, S. Ahmadi, M. Petrini, D. Aguiar, R. Mazzanti, F. Zanetto, E. Guglielmi, M. Sampietro, F. Morichetti and A. Melloni, "Control and Calibration Recipes for Photonic Integrated Circuits," in *IEEE Journal of Selected Topics in Quantum Electronics*, vol. 26, no. 5, pp. 1–10, 2020
2. F. Zanetto, A. Perino, M. Petrini, F. Toso, M. Milanizadeh, F. Morichetti, A. Melloni, G. Ferrari and M. Sampietro, "Electrical Conductance of Silicon Photonic Waveguides," in *Optics Letters*, vol. 46, no. 1, pp. 17–20, 2021
3. F. Morichetti, M. Milanizadeh, M. Petrini, F. Zanetto, G. Ferrari, D. Aguiar, E. Guglielmi, G. Ferrari, M. Sampietro and A. Melloni, "Polarization-transparent silicon photonic add-drop multiplexer with wideband hitless tuneability," in *Nature Communications*, vol. 12, 2021
4. M. Petrini, M. Milanizadeh, F. Morichetti and A. Melloni, "Dynamic Mitigation of Nonlinear Effects in Silicon Photonics Add-Drop Filter," in *Optics Letters*, vol. 46, no. 19, pp. 5023–5026, 2021
5. A. Perino, F. Zanetto, M. Petrini, F. Toso, F. Morichetti, A. Melloni, G. Ferrari and M. Sampietro, "High-sensitivity transparent photoconductors in voltage-controlled silicon waveguides," in *Optics Letters*, vol. 47, no. 6, pp. 1327–1330, 2022
6. V. Grimaldi, F. Zanetto, F. Toso, I. Roumpos, T. Chrysostomidis, A. Perino, M. Petrini, F. Morichetti, A. Melloni, N. Pleros, M. Moralis-Pegios, K. Vyrsokinos, G. Ferrari and M. Sampietro, "Self-Stabilized 50 Gb/s Silicon Photonic Microring Modulator Using a Power-Independent and Calibration-Free Control Loop", in *Journal of Lightwave Technology*, vol. 41, no. 1, pp. 218–225, 2023

7. M. Petrini, M. Seyfried, F. Morichetti and A. Melloni, "Spectral Classification and Cloning of Photonic Integrated Filters for Volume Testing," in *Journal of Lightwave Technology*, vol. 41, no. 2, pp. 645–652, 2023

8. M. Petrini, S. Seyedinnavadeh, V. Grimaldi, M. Milanizadeh, F. Zanetto, G. Ferrari, F. Morichetti and A. Melloni, "Variable Optical True Time Delay Breaking Bandwidth-Delay Constraints," in *Optics Letters*, vol. 48. no. 2, pp. 460–463, 2023

9. F. Zanetto, F. Toso, V. Grimaldi, M. Petrini, M. Milanizadeh, A. Perino, F. Morichetti, A. Melloni, G. Ferrari and M. Sampietro, "Time-multiplexed control of programmable silicon photonic circuits enabled by monolithic CMOS electronics" in *Laser and Photonics Reviews*, Manuscript Under Review

Patents

1. A. Melloni, F. Morichetti, M. Petrini, M. Milanizadeh, "Sistema e metodo di osservazione di un dispositivo ottico", WO 2022/259086-A1, filed on June 8th, 2021

Conference Proceedings

1. M. Milanizadeh, D. Aguiar, M. Petrini, E. Guglielmi, F. Zanetto, F. Toso, F. Morichetti and A. Melloni, "Automatic Tuning and Locking of Hitless Add-Drop Filters," in Group IV Photonics (GFP), Singapore, 2019

2. M. Milanizadeh, M. Petrini, F. Morichetti and A. Melloni, "Polarization insensitive tunable hitless filter for extended C band," in European Conference of Integrated Optics (ECIO), Paris (Virtual event), 2020

3. M. Petrini, M. Milanizadeh, F. Morichetti and A. Melloni, "FSR free coupled microring resonator filter on extended C-band in silicon photonics," in European Conference of Integrated Optics (ECIO), Paris (Virtual event), 2020

4. M. Milanizadeh, M. Petrini, F. Morichetti and A. Melloni, "FSR-free filter with hitless tunability across C+L telecom band," in Integrated Photonics Research, Silicon and Nanophotonics (IPR), Advanced Photonics Congress (OSA), Virtual event, 2020

5. A. Melloni, M. Milanizadeh, M. Petrini, D. Aguiar, F. Zanetto, G. Ferrari, M. Sampietro, F. Morichetti, "On the Control of Reconfigurable Photonics Integrated Circuits," (invited paper) in Italian Conference of Optics and Photonics (ICOP), Parma (Virtual Event), 2020

6. M. Petrini and M. Milanizadeh, "FSR-free Coupled Microring Resonators Hitless Photonic Filter," in Riunione Nazionale di Elettromagnetismo (RiNEm), Roma (Virtual Event), 2020

7. M. Milanizadeh, M. Petrini, F. Morichetti and A. Melloni, "Polarization-transparent FSR-free microring resonator filter with wide hitless tunability," in Optical Fiber Communication Conference (OFC), Washington D.C. (Virtual Event), 2021

8. M. Petrini, M. Milanizadeh, F. Morichetti and A. Melloni, "Automated Cloning and Lookup Table Generation for Reconfigurable Photonic Integrated Filters," in Optical Fiber Communication Conference (OFC), Washington D.C. (Virtual Event), 2021

9. M. Petrini, M. Milanizadeh, F. Morichetti and A. Melloni, "Active Compensation of Nonlinear Effects in Silicon Photonic Microring Filters," in Optical Fiber Communication Conference (OFC), Washington D.C. (Virtual Event), 2021

10. M. Petrini, M. Milanizadeh, F. Zanetto, G. Ferrari, M. Sampietro, F. Morichetti and A. Melloni, "Reconfigurable FSR-free microring resonator filter with wide hitless tunability," in IEEE Summer Topical Meeting Series (SUM 2021), Cabo San Lucas (Virtual Event), 2021

11. M. Petrini, M. Milanizadeh, F. Morichetti and A. Melloni, "Automated Lookup Table Generation and Cloning of Tuneable Photonic Integrated Filters," in Integrated Photonics Research, Silicon and Nanophotonics (IPR), Advanced Photonics Congress (OSA), Washington D.C. (Virtual Event), 2021

12. M. Petrini, M. Milanizadeh, F. Morichetti and A. Melloni, "Active Compensation of Nonlinear Distortions in Silicon Microring Resonator Filters," in Integrated Photonics Research, Silicon and Nanophotonics (IPR), Advanced Photonics Congress (OSA), Washington D.C. (Virtual Event), 2021

13. M. Petrini, M. Milanizadeh, F. Zanetto, G. Ferrari, M. Sampietro, F. Morichetti and A. Melloni, "Polarization Transparent Add-Drop Multiplexer with Hitless Tuneability," in European Conference Optical Communications (ECOC), Bordeaux, 2021

14. M. Petrini, M. Milanizadeh, F. Morichetti and A. Melloni, "Dynamic Compensation of Nonlinear Phenomena in Silicon Photonic Microring Resonator Filter," in Group IV Photonics (GFP), Malaga, 2021

15. A. Perino, F. Zanetto, M. Petrini, F. Toso, F. Morichetti, A. Melloni, G. Ferrari and M. Sampietro "Control of SiP Waveguide-Embedded Electronic Devices by Substrate/Gate Potential Tuning," in Group IV Photonics (GFP), Malaga, 2021

16. M. Petrini, R. Baldi, M. Seyfried, F. Morichetti and A. Melloni, "Automatic Testing of Silicon Photonic Add/Drop Multiplexer," in European Conference on Integrated Optics (ECIO), Milan, 2022

17. A. Perino, F. Zanetto, M. Petrini, F. Morichetti, A. Melloni, G. Ferrari and M. Sampietro, "SiP Waveguide-Embedded Electronic Devices controlled by Substrate/Gate Potential Tuning," in European Conference on Integrated Optics (ECIO), Milan, 2022

18. F. Zanetto, F. Toso, V. Grimaldi, M. Petrini, M. Milanizadeh, A. Perino, P. Piedimonte, F. Morichetti, A. Melloni, G. Ferrari and M. Sampietro, "Monolithically Integrated Electronics in Zero-Change Silicon Photonics," in European Conference on Integrated Optics (ECIO), Milan, 2022

19. C. De Vita, A. Perino, F. Zanetto, F. Toso, G. Ferrari, M. Petrini, A. Melloni, N. G. Pruiti, M. Sorel and F. Morichetti "Light monitoring in silicon-based photonic integrated platforms for near-infrared and visible light applications," in International Conference on Optical, Optoelectronic and Photonic Materials and Applications (ICOOPMA), Ghent, 2022

20. M. Petrini, R. Baldi, F. Morichetti and A. Melloni, "Automatic Testing of a Silicon Photonic Reconfigurable Add/Drop Multiplexer," in IEEE Summer Topical Meetings (SUM), Cabo San Lucas, 2022, **Invited Paper**

21. M. Petrini, R. Baldi, M. Seyfried, F. Morichetti and A. Melloni, "Probe-Card Based Automatic Testing of Photonic Integrated Filters," in Integrated Photonics Research, Silicon and Nanophotonics (IPR), Advanced Photonics Congress (OSA), Maastricht, 2022

22. M. Petrini, S. Seyedinnavadeh, F. Morichetti and A. Melloni, "Breaking the Delay-Bandwidth Product of Continuously-Tunable MZI Delay-Line," in Integrated Photonics Research, Silicon and Nanophotonics (IPR), Advanced Photonics Congress (OSA), Maastricht, 2022

23. M. Petrini, S. Seyedinnavadeh, F. Morichetti and A. Melloni, "Continuously Tuneable MZI-based Delay Line Overcoming Delay-Bandwidth Product," in European Conference Optical Communications (ECOC), Basel, 2022

Contents

Acronyms

ADC	Analog to Digital Converter
AI	Analog Input
AO	Analog Output
ASE	Amplified Spontaneous Emission
AWG	Arrayed Waveguide Gratings
BBS	Broadband Source
BER	Bit Error Rate
BJTs	Bipolar Junction Transistors
BPF	Band Pass Filter
CD	Chromatic Dispersion
CMOS	Complementary Metal Oxide Semiconductor
CROW	Coupled-Resonators Optical Waveguide
DAC	Digital to Analog Converter
DFT	Design For Testability
DMM	Digital Multimeter
DP-16QAM	Double Polarization-Quadrature Amplitude Modulation
DP-QPSK	Double Polarization-Quadrature Phase Shift Keying
DSP	Digital Signal Processor
DUT	Device Under Test
EDFA	Erbium Doped Fiber Amplifier
EIC	Electronic Integrated Circuit
ER	Extinction Ratio
FCA	Free Carrier Absorption
FCD	Free Carrier Dispersion
FDTD	Finite Difference Time Domain
FEC	Forward Error Correction
FET	Field Effect Transistors
FFC	Flat Flex Cable

FIR	Finite Impulse Response
FPGAs	Field Programmable Gated Arrays
FSR	Free Spectral Range
FWM	Four Wave Mixing
I/O	Input-Output
IFFT	Inverse Fast Fourier Transform
IIR	Infinite Impulse Response
INA	Instrumentation Amplifier
ISI	Intersymbol Interference
LUT	LookUp Table
MEMS	Micro-Electro-Mechanical Systems
MFD	Mode Field Diameter
MRR	Microring Resonator
MSE	Mean Squared Error
MUX	Multiplexer
MZI	Mach-Zehnder Interferometer
NRZ	Non-Return to Zero
OA	Operational Amplifier
OCR	Optical Character Recognition
OOK	On-Off Keying
OSA	Optical Spectrum Analyzer
OSNR	Optical Signal to Noise Ratio
OTDM	Optical Time Domain Multiplexing
PB	Passband
PC	Probe Card
PCB	Printed Circuit Board
PD	Photodetector
PDL	Polarization-Dependent Loss
PI	Proportional Integral
PIC	Photonic Integrated Circuit
PID	Proportional Integral Derivative
PM	Power Meter
PMD	Polarization Mode Dispersion
PRC	Polarization Rotator and Combiner
PS	Polarization Scrambler
PSD	Power Spectral Density
PSR	Polarization Splitter and Rotator
REF	Reference
SB	Stopband
SD-FEC	Soft Decision Forward Error Correction
SMU	Source and Measurement Unit

SOI	Silicon-On-Insulator
SOP	State of Polarization
SPM	Self-Phase Modulation
TC	Tuneable Coupler
TDM	Time Domain Multiplexing
TEC	Thermo-Electric Cooler
TED	Thermal Eigenmode Decomposition
TIA	Trans-Impedance Amplifier
TiN	Titanium Nitride
TLS	Tuneable Laser Source
TOADM	Tuneable Optical Add/Drop Multiplexer
TPA	Two Photon Absorption
VNA	Vectorial Network Analyzer
VOA	Variable Optical Attenuator
WDL	Wavelength-Dependent Loss
WDM	Wavelength Division Multiplexing
WLT	Wafer Level Testing
XPM	Cross-Phase Modulation

List of Figures

List of Tables

1.1   Photonic Integrated Circuits

Photonic Integration is considered one of the enabling technologies for solutions to the challenges of the next decades.

Traditionally, Optical Communications, characterized by high data rates and transmission over longer and longer distances, have always been the driving market, steering the development of Photonics. Integrated optical circuits, in fact, show appealing features for data- and tele- communication purposes [1], such as their low power consumption and extremely large bandwidth, at a relatively low cost [2, 3].

Today, other fields are also experiencing these benefits, such as 5G networks [4, 5], optical interconnects [6–8], sensing [9] and bio-sensing [10, 11], quantum communication [12] and computation [13]. And we expect the development of novel applications in the next few years, related for example to Artificial Intelligence [14, 15] or Cryptocurrencies [16].

Although Photonics has been studied since early 1970s', optical components cannot fill most of the market requests.

However, thanks to the continuous improvement of the technological processes in commercial foundries and the advance of knowledge, expertise and pervasiveness, PICs are increasing their integration density and complexity.

In the recent years, two technologies have become dominant: Indium Phosphide [17] and Silicon [18–20]. Both of them, having their own pros and cons, are semiconductor based on high index contrast, offering the possibility of integrating many sub-parts on the same chip.

For example, the former allows the monolithic integration of light sources, detectors and waveguides (even though it is a quite expensive technology). The latter, instead, shows a higher density, in terms of building blocks per mm^2, compatibility with the CMOS-process and lower cost per unit area [21].

In this work we mainly focus on Silicon Photonic platforms.

© The Author(s), under exclusive license to Springer Nature Switzerland AG 2025   1
M. Petrini, *Mixed-Signal Generic Testing in Photonic Integration*, Synthesis Lectures
on Digital Circuits & Systems, https://doi.org/10.1007/978-3-031-60811-7_1

Exploiting this technology and the possibility of miniaturization that it offers, several interesting PICs with a high degree of complexity have been proposed and successfully validated [22, 23]. Furthermore, thanks to improvements that have been demonstrated in terms of integration (which must be separated from miniaturization [24]), new functionalities can be embedded in a single photonic chip. Then, by merging photonics with (well known) electronics, these functions meet the requirements of reconfigurability and adaptability [25, 26]. Thus, in Photonics, we are observing the transition from "stand-alone-device" to "system-on-chip".

This shift unavoidably makes production, validation, characterization and testing much more complex, but at the same time enables new possibilities and unprecedented solutions.

1.2   Photonic Testing

Photonics is becoming increasingly widespread, penetrating in many different fields. Technological processes appear to be quite mature and complex devices (i.e., with several building blocks per mm^2) can be realized and their functionalities have been demonstrated.

However, we are still far from making Photonics a mass-production technology. There are many reasons behind this statement, and one of them is the total cost of the last stages of the supply chain (i.e., assembly, testing and packaging), that turn a die into a final product.

These steps, in total, represent more than 80% of the total cost of a Photonic Device. In particular, "testing" contribution is $\approx$ 30% (Fig. 1.1a). Such an enormous impact is related to two main aspects. First of all, photonic testing shows a twofold nature. A photonic component must be validated, in fact, from the electrical and optical standpoints. This dramatically increases the complexity, resource and time consumption of PIC testing, with respect to, for example, testing in electronics or microwave domains. Furthermore, if compared with these two disciplines, Photonics is still young and unripe. In many steps of the production chain, and especially in testing, there is an evident lack of standardization (even though, in this regard, some solutions have been recently proposed [27]), which turns into rising costs in a scale economy.

In general, and Photonics is not an exception, the term "testing" identifies a class of operation aimed at the qualification, validation and verification of a product.

Typically, test engineers identify two approaches [30]:

- black-box testing (or functional testing),
- white-box testing (or structural testing).

These two strategies are not mutually exclusive and often they are executed together, even if (or because of) they are intrinsically different.

Functional testing is a "blind" technique to perform product assessment. In principle, the behaviour of the DUT can be checked, without any information about its technology,

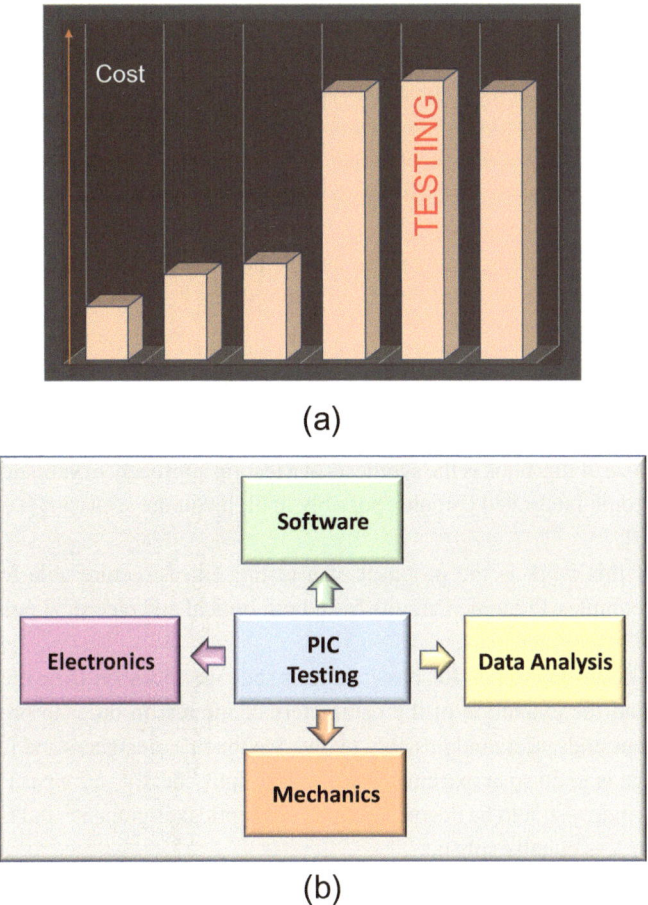

Fig. 1.1 **a** Impact of the testing in total cost of a packaged PIC [27–29]. **b** The different disciplines related to photonic testing.

topology or implementation. The DUT is a black box and the test engineer ignores its content. The main aspects to be taken into account are the input provided by the testing system, the expected output and the real output. Evaluating the differences between the ground truth and the expected outcomes, the DUT can be classified and validated. This kind of testing is usually faster and more flexible than the other one, however its outcome is deprived of any clue on the possible failures of the DUTs. This approach can be sufficient for "simple" devices, with a limited number of degrees of freedom (such as dummy light sources, single cavities or single interferometers).

When, instead, complex photonic circuits, composed of several sub-parts, are considered, testing cannot be separated from calibration or programming processes, which become part of the validation stage itself. In fact, for many applications, the time/frequency domain functionalities have to be evaluated, but due to unavoidable fabrication imperfections, it is impossible to assess PIC behaviour on "as built" devices.

Thus, structural testing (which means that the structure of the DUT is known and/or it can be mathematically and univocally modeled) is needed. This implies that testing activities become heavier, slower and less flexible, but its outcome provides hints on PIC failures.

Regardless of the strategy adopted to perform photonic testing, it is worth mentioning that this process is very demanding [31], and extremely multidisciplinary (Fig. 1.1b).

As we have seen, photonic testing presents several challenges that have to be approached and solutions will be provided and discussed through this book.

1.3   Book Purposes

The main purpose of this book is the synthesis of a testing approach, to validate the behaviour of Silicon Photonic Integrated Circuits, possibly in high volume scenarios [i.e., Wafer Level Testing (WLT)].

The aim of this work is the proposal of a testing infrastructure able to test the new generation of complex Photonic Circuits from both optical and electrical points of view, in a fast, efficient and reliable way.

We do not describe recipes for pass/fail tests, but we focus on their electrical/optical performance and the extraction of the parameters of interest in order to properly classify the set of devices under test (and possibly to give feedback to designers and foundries). The actual challenge is to do so at extremely high throughput, ideally < 2 s per DUT.[1]

This time window should be devoted to electrical/optical alignment[2] and to the measurements (more or less equally split).

Furthermore, the duty is even tougher considering the complexity of the DUTs and their sensitivity to the fabrication imperfections. An error of just 1 nm (two Silicon atoms) in the waveguide dimensions (width or height), during the fabrication process, leads to a non-negligible shift (i.e., > 100 GHz) in the spectral response of the DUT (if spectrally selective) [32, 33]. If the considered PIC is fairly complex (i.e., constituted by several building blocks) it would be impossible test its optical functionalities (in terms of transfer function), as it comes "out-of-fab".

---

[1] Considering tens of thousands of devices in a 12—inches wafer, the entire test lasts only a few hours.

[2] Commercial robots can perform probe landing and optical alignment (movement between two consecutive devices) in a sub-second time scale (see https://www.ficontec.com), while in a research laboratory these tasks were performed in approximately 3 s, by using semi-automatic Melles-Griot micropositioners.

For this reason, a reliable and repeatable technique has been developed, in order to:

1. state the (spectral) differences between DUT and a certain reference;
2. use the information in the previous point to tune the DUT (if necessary), to check its optical functionalities and eventually classify it.

To accomplish these two tasks, a suitable electrical test controller was assembled and its analog front-end was carefully designed.

This infrastructure is capable of performing a pure electrical testing and of executing the calibration algorithm, within the requirements, in terms of speed and repeatability.

Once the tuning is completed, and the specific DUT passes the test (from both electrical and optical standpoints), the working point can be stored in a proper memory structure, called LUT. In so doing, during normal operation, the desired state of the device can be recalled, without performing any other tuning procedure. We believe that the calibration and LUT filling should be part of the testing at the wafer scale (to do so mechanical systems for optical and electrical alignment should be extremely reliable). For tuneable/programmable PICs, the only way to guarantee that their behaviour (in time and/or frequency domains) is in line with specifications. There could be unlucky devices that may not reach the converged state (for example due to a more lossy waveguide or due to a non-tuneable directional coupler far away from the condition it is designed for), and thus they have to be discarded. This can be conveniently done before packaging, in order to save time and resources (discarding an already packaged device would not be very practical and extremely expensive).

1.4   Original Contribution

We believe that the outcomes reported in this work could have a beneficial impact in the Integrated Photonics field, proposing (hopefully) some interesting solutions for volume testing.

- **Novel topologies of Photonic Integrated Circuits.**
  In Chaps. 2 and 6, two novel PICs are presented and discussed.
  Indeed their design has been a team work, that involves a number of actors. The main contribution reported here is the implementation of a numerical simulator (at the circuit level), which is useful for predicting their time/frequency behaviour based on different choices of their degrees of freedom [i.e., couplers, rings' radii, imbalances of the Mach-Zehnder Interferometers].
  Specifically, in Chap. 2 design and characterization of a four-channel-TOADM are described. Each channel of this device is a 4th order-Microring Resonator-based filter, having appealing features, never shown all together in the same device. Among those, aperiodic spectral response, hitless tuning and polarization insensitivity are worth men-

tioning. The simultaneous implementation of these characteristics makes the whole PIC quite complex. Thus, it becomes the perfect benchmark for the validation of the testing techniques.

In Chap. 6, instead, we go through the two continuously tuneable delay lines implemented by means of Mach-Zehnder Interferometers, cascaded or in a nested configuration. These topologies overcome the traditional limitations of delaying structures, namely the bandwidth-delay product constraint. A testing procedure is also synthesized for these PICs.

These two families of photonic circuits (filters and delay lines), sharing the same platform -SOI- and having the same actuators and sensors, can be tested exploiting the same electronic infrastructure.

- **Synthesis of a electro-optical testing technique.**
  Chapters 3 and 4 describe all the steps to follow, in order to test (in a sub-second scale) a complex photonic circuit, from the both electrical and optical sides. Single ingredients involved in this recipe are very well known and established (tuning algorithm, testing/-control electronics, electrical PC and so on), but here we put all the bricks together and we synthesized novel testing techniques, giving a comprehensive outlook.

- **Estimation of Thermal Crosstalk effects.**
  In Chap. 3, we discuss the estimation of the thermal coupling among the different building blocks of the PIC under test. We do not rely on heuristic approaches or on optical measurements, but we exploit the thermo-optic actuators (already embedded in the DUT) as thermal probes. The approach is not new by itself, but here the so called crosstalk matrix ($\mathbf{T}$) becomes an outcome of testing activity. This information is crucial for the following steps of the photonic testing (pre-calibration and optical classification) and for the execution of other [thermal crosstalk free, for example based on Thermal Eigenmode Decomposition (TED)] calibration loops during the normal operation.

- **Cloning Technique.**
  In Chap. 4, we calibrate (under certain conditions) a programmable PIC (DUT) in such a way that its spectral response replicates that of the "golden reference" device. Basically, to tune the DUT, we propose the use of a signal, which is obtained by cascading a Broadband Source (BBS) and the reference device. In the literature, the use of custom Power Spectral Density (PSD) to tune a filter has been proposed and effectively demonstrated. However, this approach has never been used (at least, in a testing scenario) to evaluate differences between DUT and a specific reference. Moreover, this information has never been used to perform "calibration for testing" and hence classification of PICs.

- **Automatic LookUp Table Generation.**
  Once the PIC passes the test (from both electrical and optical standpoints) and calibration reaches convergence, the working points of its actuators are stored in LUTs. These memory structures (very well known in other fields of information technology, such as digital electronics) are dynamic, and they can be updated upon different user needs (as it occurs

in the case of programmable photonic circuits) or in case of environmental conditions change (e.g., temperature drifts).

LUTs are the true outcome of photonic testing, and they are also crucial during normal operation of the PIC, since a certain state can be easily recovered. Thus, we believe that the intrinsic value of a photonic circuit is dramatically enhanced by these tables.

- **Dynamic Mitigation of Nonlinear Effects.**
  In Chap. 5 we use the same technique used to perform cloning, to counteract (from the spectral point of view) nonlinear effects on the Add/Drop filter. Usually, nonlinearities in Silicon Photonics have very detrimental effects and they can be overcome recurring to quite complicated solutions, that rely on the implementation of athermal waveguides and/or substantial modification of nanowire geometry. Here, instead, we use a simple control loop, that acts locally (i.e., only on the building blocks suffering the effects of nonlinear phenomena) and, in presence of a high-power input signal, optimizes the spectral shape of the filter, recovering its relevant parameters (such as return loss and bandwidth). When the optimum is reached, the LUT can be extended, including these new working points, for this extreme condition. Then, during normal operation, these values can be recalled and the filter can be pre-distorted and ready to accept high-intensity input signals.

1.5 Book Overview

The book is organized as follows.

- **Chapter 2: The Device Under Test.**
  In this chapter the design of the DUT—namely, a TOADM—is described in detail, starting from its own building blocks. Most appealing features of this PIC are presented. Then, the fabricated device is characterized, at both the device and system levels, validating all its key aspects. This chapter concludes with some considerations about fabrication imperfections, and their impact on PIC features. This chapter aimes to present the expected specifications of the DUT and to understand its working principle. A priori information about the DUT always enhance testing performance.
- **Chapter 3: Electrical Testing.**
  Since DUT -for programmability and calibration purposes- embeds electrical devices (thermo-optic actuators and sensors), an electrical testing is definitely needed. In this chapter we propose a flexible infrastructure capable of testing the DUT from an electrical point of view (and of executing simple control algorithms). This is composed of commercial NI-PXI controller, with its own multichannel Analog to Digital Converter (ADC) and Digital to Analog Converter (DAC), a custom analog board (for testing signal conditioning and recovery) and an electrical PC. The problem of (electrical and optical)

coupling is addressed and the PC is analyzed. Finally, the electrical testing outcome is discussed and a method to estimate the thermal crosstalk is proposed.

- **Chapter** 4: **Techniques and Methods for Optical Testing**.
  In this Chapter we introduce a technique to perform classification and "calibration for testing" of spectrally selective PICs. In this way, we can state not only how far an "out-of-fab" DUT is from the reference state, but also we can use this information tune it, in order to resemble the reference one (from the frequency response standpoint). The approach is validated by means of proper numerical simulations. Then, the technique is applied to the actual DUT and its performance in terms of throughput (i.e., number of PICs tuned/classified per unit time) are stated. A posteriori spectral measurements show its effectiveness and reliability. The method is executed exploiting the electronics described in the previous chapter.

- **Chapter** 5: **Mitigation of Nonlinear Effects**.
  In this chapter we show how to update the LUT in presence of high-power signals, causing the manifestation of nonlinear phenomena on the described DUT. These, in fact, are triggered at a low threshold and strongly limit the use of Silicon Photonics. After discussing the effect of nonlinearities, at both device and system levels, we show the use of the tuning algorithm investigated in Chap. 4, to counteract nonlinear effects. The benefit of the solution is demonstrated by means of spectral measurements and system level assessments. In the last paragraph of the Chapter we report on the LUT update.

- **Chapter** 6: **Testing of Delay Lines Breaking Bandwidth-Delay Constraint**.
  The last Chapter of the book is dedicated to the description of two novel true time optical delay lines, overcoming the traditional limitations (related to Bandwidth-Delay product), made in Silicon Photonics. After discussing their topology, design and principle of operation, we introduce a calibration technique, executed by the aforementioned electronics. Using these two ingredients we can cook a testing recipe, to validate the behaviour of the two different delay lines under test. A posteriori measurements show the effectiveness of the method.

References

1. J. Doylend and A. Knights, "The evolution of silicon photonics as an enabling technology for optical interconnection," *Laser and Photonics Reviews*, vol. 6, no. 4, 2012.
2. D. A. B. Miller, "Reconfigurable add-drop multiplexer for spatial modes," *Optics Express*, vol. 21, no. 17, 2013.
3. C. R. Doerr, "Proposed Architecture for MIMO Optical Demultiplexing Using Photonic Integration," *IEEE Photonics Technology Letters*, vol. 23, no. 21, 2011.
4. J. Capmany and P. Munoz, "Integrated Microwave Photonics for Radio Access Networks," *J. Lightwave Technol.*, vol. 32, no. 16, 2014.
5. R. Waterhouse and D. Novack, "Realizing 5G: Microwave Photonics for 5G Mobile Wireless Systems," *IEEE Microwave Magazine*, vol. 16, no. 8, 2015.

6. T. Barwicz, H. Byun, F. Gan, C. W. Holzwarth, M. A. Popovic, P. T. Rakich, M. R. Watts, E. P. Ippen, F. X. Kaertner, H. I. Smith, J. S. Orcutt, R. J. Ram, V. Stojanovic, O. O. Olubuyide, J. L. Hoyt, S. Spector, M. Geis, M. Grein, T. Lyszczarz, and J. U. Yoon, "Silicon photonics for compact, energy-efficient interconnects (Invited)," *Journal of Optical Networking*, vol. 6, no. 1, 2007.

7. P. Chaisakul, D. Marris-Morini, J. Frigerio, D. Chrastina, M.-S. Rouifed, S. Cecchi, P. Crozat, G. Isella, and L. Vivien, "Integrated germanium optical interconnects on silicon substrates," *Nature Photonics*, vol. 8, no. 6, 2014.

8. D. A. B. Miller, "Optical interconnects to electronic chips," *Applied Optics*, vol. 49, no. 25, 2010.

9. W. Xie, T. Komljenovic, J. Huang, M. Tran, M. Davenport, A. Torres, P. Pintus, and J. Bowers, "Heterogeneous silicon photonics sensing for autonomous cars (Invited)," *Optics Express*, vol. 27, no. 3, 2019.

10. C. Ciminelli, F. Dell'Olio, D. Conteduca, and M. N. Armenise, "Silicon photonic biosensors," *IET Optoelectronics*, vol. 13, no. 2, 2019.

11. P. Borga, F. Milesi, N. Peserico, C. Groppi, F. Damin, L. Sola, P. Piedimonte, A. Fincato, M. Sampietro, M. Chiari, A. Melloni, and R. Bertacco, "Active Opto-Magnetic Biosensing with Silicon Microring Resonators," *Sensors*, vol. 22, no. 9, 2022.

12. M. Foertsch and S. Hengesbach, "Enabling Quantum Technologies with Photonics," in *Conference on Lasers and Electro-Optics*, Optica Publishing Group, 2021.

13. J. W. Silverstone, D. Bonneau, K. Ohira, N. Suzuki, H. Yoshida, N. Iizuka, M. Ezaki, C. M. Natarajan, M. G. Tanner, R. H. Hadfield, V. Zwiller, G. D. Marshall, J. G. Rarity, J. L. O'Brien, and M. G. Thompson, "On-chip quantum interference between silicon photon-pair sources," *Nature Photonics*, vol. 8, no. 2, 2013.

14. M. Miscuglio and V. J. Sorger, "Photonic tensor cores for machine learning," *Applied Physics Reviews*, vol. 7, no. 3, 2020.

15. H.-T. Peng, J. C. Lederman, L. Xu, T. F. de Lima, C. Huang, B. J. Shastri, D. Rosenbluth, and P. R. Prucnal, "A Photonics-Inspired Compact Network: Toward Real-Time AI Processing in Communication Systems," *IEEE Journal of Selected Topics in Quantum Electronics*, vol. 28, no. 4, 2022.

16. S. Pai, T. Park, B. Penkovsky, M. Milanizadeh, M. Ball, M. Dubrovsky, N. Abebe, F. Morichetti, A. Melloni, O. Solgaard, and D. A. Miller, "LightHash: Experimental Evaluation of a Photonic Cryptocurrency," in *Conference on Lasers and Electro-Optics*, Optica Publishing Group, 2022.

17. M. Smit, X. Leijtens, H. Ambrosius, E. Bente, J. van der Tol, B. Smalbrugge, T. de Vries, E.-J. Geluk, J. Bolk, R. van Veldhoven, L. Augustin, P. Thijs, D. D'Agostino, H. Rabbani, K. Lawniczuk, S. Stopinski, S. Tahvili, A. Corradi, E. Kleijn, D. Dzibrou, M. Felicetti, E. Bitincka, V. Moskalenko, J. Zhao, R. Santos, G. Gilardi, W. Yao, K. Williams, P. Stabile, P. Kuindersma, J. Pello, S. Bhat, Y. Jiao, D. Heiss, G. Roelkens, M. Wale, P. Firth, F. Soares, N. Grote, M. Schell, H. Debregeas, M. Achouche, J.-L. Gentner, A. Bakker, T. Korthorst, D. Gallagher, A. Dabbs, A. Melloni, F. Morichetti, D. Melati, A. Wonfor, R. Penty, R. Broeke, B. Musk, and D. Robbins, "An introduction to InP-based generic integration technology," *Semiconductor Science and Technology*, vol. 29, no. 8, 2014.

18. T. Baehr-Jones, T. Pinguet, P. L. Guo-Qiang, S. Danziger, D. Prather, and M. Hochberg, "Myths and rumours of silicon photonics," *Nature Photonics*, vol. 6, no. 4, 2012.

19. A. Rahim, T. Spuesens, R. Baets, and W. Bogaerts, "Open-Access Silicon Photonics: Current Status and Emerging Initiatives," *Proceedings of the IEEE*, vol. 106, no. 12, 2018.

20. D. Thomson, A. Zilkie, J. E. Bowers, T. Komljenovic, G. T. Reed, L. Vivien, D. Marris-Morini, E. Cassan, L. Virot, J.-M. Fédéli, J.-M. Hartmann, J. H. Schmid, D.-X. Xu, F. Boeuf, P. O'Brien, G. Z. Mashanovich, and M. Nedeljkovic, "Roadmap on silicon photonics," *Journal of Optics*, vol. 18, no. 7, 2016.

21. C. Kopp, S. Bernabé, B. B. Bakir, J. Fedeli, R. Orobtchouk, F. Schrank, H. Porte, L. Zimmermann, and T. Tekin, "Silicon Photonic Circuits: On-CMOS Integration, Fiber Optical Coupling, and Packaging," *IEEE Journal of Selected Topics in Quantum Electronics*, vol. 17, no. 3, 2011.
22. P. Dumais, D. J. Goodwill, D. Celo, J. Jiang, C. Zhang, F. Zhao, X. Tu, C. Zhang, S. Yan, J. He, M. Li, W. Liu, Y. Wei, D. Geng, H. Mehrvar, and E. Bernier, "Silicon Photonic Switch Subsystem With 900 Monolithically Integrated Calibration Photodiodes and 64-Fiber Package," *Journal of Lightwave Technology*, vol. 36, no. 2, 2018.
23. D. Pérez, I. Gasulla, L. Crudgington, D. J. Thomson, A. Z. Khokhar, K. Li, W. Cao, G. Z. Mashanovich, and J. Capmany, "Multipurpose silicon photonics signal processor core," *Nature Communications*, vol. 8, no. 1, 2017.
24. R. Chau, B. Doyle, S. Datta, J. Kavalieros, and K. Zhang, "Integrated nanoelectronics for the future," *Nature Materials*, vol. 6, no. 11, 2007.
25. M. Milanizadeh, S. SeyedinNavadeh, F. Zanetto, V. Grimaldi, C. D. Vita, C. Klitis, M. Sorel, G. Ferrari, D. A. B. Miller, A. Melloni, and F. Morichetti, "Separating arbitrary free-space beams with an integrated photonic processor," *Light: Science and Applications*, vol. 11, no. 1, 2022.
26. D. A. B. Miller, "Self-configuring universal linear optical component (Invited)," *Photonic Research*, vol. 1, no. 1, 2013.
27. S. Latkowski, D. Pustakhod, M. Chatzimichailidis, W. Yao, and X. J. M. Leijtens, "Open Standards for Automation of Testing of Photonic Integrated Circuits," *IEEE Journal of Selected Topics in Quantum Electronics*, vol. 25, no. 5, 2019.
28. "AIM Photonics Academy, 2020-Integrated Photonics System Roadmap International (IPSR-I)," 2020.
29. E. Fuchs, E. Bruce, R. Ram, and R. Kirchain, "Process-based cost modeling of photonics manufacture: the cost competitiveness of monolithic integration of a 1550-nm DFB laser and an electroabsorptive modulator on an InP platform," *Journal of Lightwave Technology*, vol. 24, no. 8, 2006.
30. P. Reichert, "System Level Test (White Paper)," *Teradyne*, 2022.
31. R. Polster, L. Y. Dai, Q. Cheng, M. Oikonomou, S. Rumley, and K. Bergman, "Challenges and solutions for high-volume testing of silicon photonics," in *Silicon Photonics XIII* (G. T. Reed and A. P. Knights, eds.), SPIE, 2018.
32. Z. Lu, J. Jhoja, J. Klein, X. Wang, A. Liu, J. Flueckiger, J. Pond, and L. Chrostowski, "Performance prediction for silicon photonics integrated circuits with layout-dependent correlated manufacturing variability," *Optics Express*, vol. 25, no. 9, 2017.
33. T.-H. Yen and Y.-J. Hung, "Fabrication-Tolerant CWDM (de)Multiplexer Based on Cascaded Mach–Zehnder Interferometers on Silicon-on-Insulator," *Journal of Lightwave Technology*, vol. 39, no. 1, 2021.

The Device Under Test

2

2.1 Introduction

In this chapter the PIC used as a DUT (for what concerns Chaps. 3, 4 and 5) is introduced, its architecture is deeply investigated and its operation is described.

In this work, we examine a Silicon Photonic TOADM, one of the key elements to manage the evolution of new generation core/backhaul networks, and inter/intra datacenter interconnects [1, 2].

Currently, the dynamic traffic allocation, the light-path routing flexibility and the network mesh reconfiguration in an efficient, fast and automated way, are paramount features, given the continuously growing capacity demand and the necessity of shorter and shorter latency [3–5].

In particular, the proposed PIC, which fulfils most of these needs, not only has a high impact on the next generation optical networks, but is also complex enough to be used as a case study for the development and validation of new testing techniques, as discussed in the following sections and chapters.

This chapter is organized as follows:

- In Sect. 2.2 a general overview of TOADM is provided, including a list of its own building blocks.
- In Sect. 2.3 the design of the single Add/Drop channel (namely an Add/Drop filter) of the multiplexer is described.
- In Sect. 2.4 the results of the deep characterization carried out on the device are reported.
- In Sect. 2.5 the impact of the fabrication tolerances on the filter spectral responses is analyzed, by means of suitable numerical simulations.

© The Author(s), under exclusive license to Springer Nature Switzerland AG 2025   11
M. Petrini, *Mixed-Signal Generic Testing in Photonic Integration*, Synthesis Lectures
on Digital Circuits & Systems, https://doi.org/10.1007/978-3-031-60811-7_2

2.2   Tuneable Optical Add/Drop Multiplexer

In this section the architecture of the optical multiplexer is described. Its topology is sketched in Fig. 2.1a, while a top-view microphotograph is reported in Fig. 2.1b. The footprint is 5 mm by 2.7 mm.[1]

It can insert/drop up to four different wavelength channels into/from the bus waveguide -and thus into/from the Wavelength Division Multiplexing (WDM) network- in a reconfigurable manner.

In the recent decades, Silicon Photonics has been considered very promising for the implementation of this kind of WDM architectures, given its low cost, low power consumption, small footprint and Complementary Metal Oxide Semiconductor (CMOS) processes compatibility [7, 8].

Thus, the device (designed with Nazca library [9]) has been realized in a commercial 220 nm-silicon-photonic platform [10].

The enabling building block of such a device is the single Add/Drop filter, implemented by means of directly coupled Microring Resonators. Ring resonators-based devices, made in high-index contrast platforms, offer spectral features compliant with those typically needed by an optical Add/Drop multiplexer (in a relatively compact footprint), such as wide-bandwidth passbands (>10 GHz), high extinction ratio (>50 dB) and quite steep roll-off [11–13].

Although the design and structure of the single filter will be further clarified in the next section, we can state that the proposed device simultaneously fulfills three main requirements of the next generation optical networks, such as:

● ultra-wide operational wavelength range (i.e., wider than the C-band),

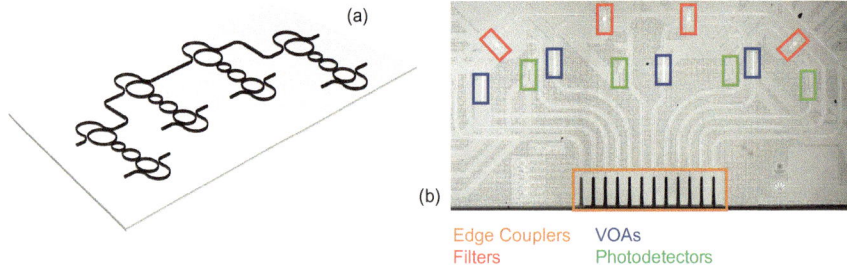

Fig. 2.1 **a** Topology of the Tuneable Optical Add/Drop Multiplexer (TOADM), used as Device under Test (DUT) through the book and **b** top-view microphotograph of the fabricated device. Adapted from [6], with permission

---

[1] The footprint is mostly dominated by the presence of 52 electrical pads, all around the photonic multiplexer, whose area is 90 μm by 90 μm. Further comments on this will be provided in the next sections and chapters.

- hitless tuneability (i.e., dynamic reconfiguration of the multiplexer, without impairing the transmission of the channels propagating through the device),
- polarization transparency, to exploit the maximum capacity of the network.

These features have been previously demonstrated [12, 14–17], but never at the same time on the same device.

Nevertheless, alongside the single filter, other building blocks [10] play a key role in the good operation of this system-on-chip, such as monitor Germanium Photodetectors [shown in detail, in Fig. 2.2a], placed at one of the outputs of each filter and suspended couplers. These mode adapters, whose microphotograph is shown in Fig. 2.2b, ensure a highly efficient fiber-to-chip Input-Output (I/O) coupling.

Moreover, the whole device presents ten optical I/Os and fifty-two electrical pads, which are useful for reading every single Photodetector (PD) and to drive actuators (heaters and p-i-n junctions) embedded in the photonic circuit.

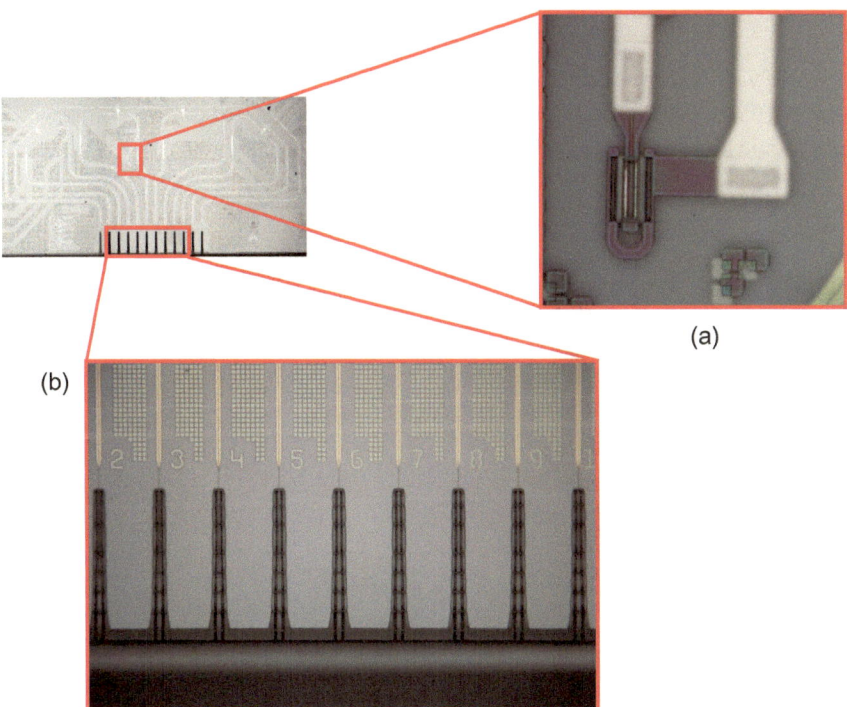

(a)

(b)

Fig. 2.2 **a** Top-view microphotograph of the Ge-photodetector placed at the Drop output of the single TOADM channel. **b** Top-view microphotograph of the suspended couplers (mode adapters) used as optical Inputs/Outputs

2.3   Silicon Photonic Add/Drop Filter Design

The single filter topology and its top-view microphotograph are reported in Fig. 2.3a and b, respectively. This building block (whose footprint is 170 μm by 30 μm) consists of a directly coupled series of ring resonators, connected to two bus waveguides by means of Tuneable Couplers (realized using Mach-Zehnder Interferometers). In the inset of Fig. 2.3a, the cross section of the employed nanowire is shown. It is rib-shaped, characterized by a nominal width of 500 nm and a height of 220 nm, while the lateral slab has a thickness of 90 nm. To fully reconfigure the filter, each Microring Resonator (MRR) phase and each Tuneable Coupler (TC) working point can be individually controlled by a TiN micro-heaters deposited 700 nm above the waveguides. P-i-n junctions [surrounding the two central rings, as shown in Fig. 2.3c] also aid reconfigurability.

In the following, the most appealing features of this photonic filter are presented.

2.3.1   FSR-Free Filter

To effectively cover a large operative wavelength range [i.e., Free Spectral Range Free Spectral Range (FSR) >4 THz], as would have needed for this kind of devices, rings with radii <3 μm should be adopted. However, this is unfeasible since it leads to a severe performance

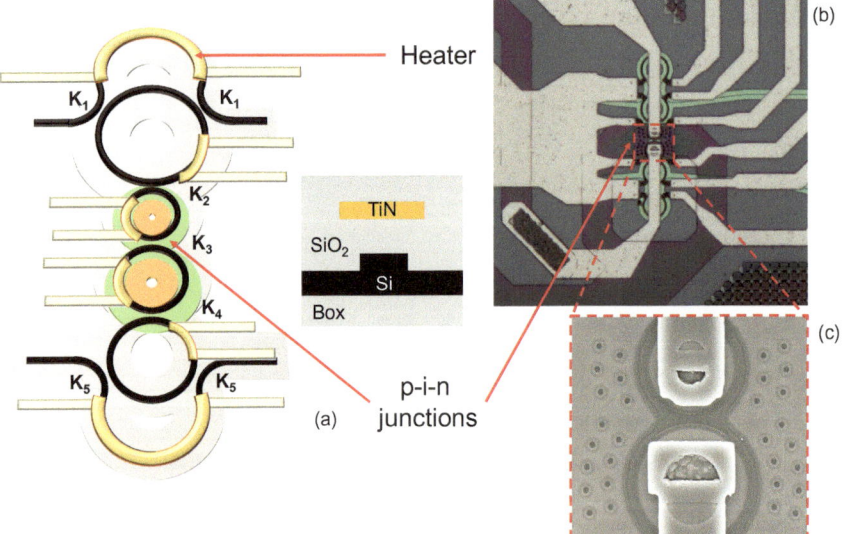

Fig. 2.3 a Topology of the single 4th order MRR filter, including heaters and p-i-n junctions. In the inset the cross-section of the exploited waveguide is reported. **b** Top-view microphotograph of the filter. **c** Detail of the p-i-n junctions. Adapted from [6], with permission

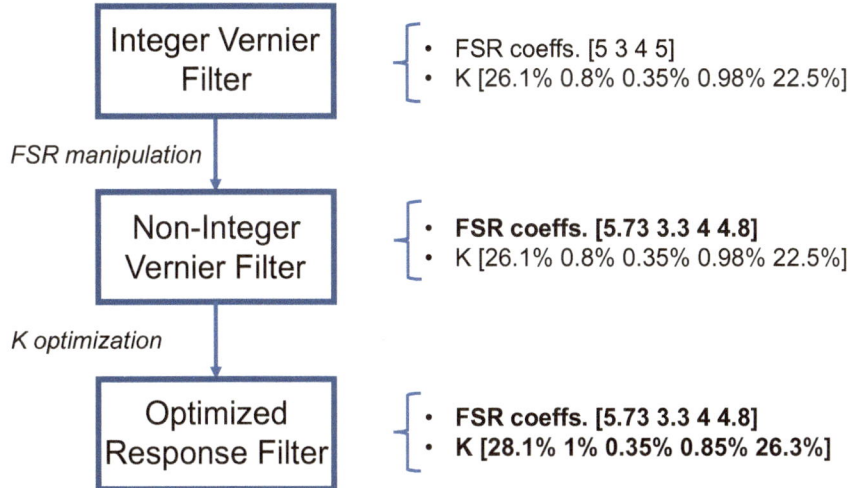

Fig. 2.4 Flowchart of the steps executed during the design of the MRR filter

degradation (in particular high curvature-induced radiation). Furthermore, in case of thermo-optic tuning (as in our case and we will investigate it in the following) the size of a single MRR is also limited by the maximum temperature achievable by the heater, controlling its phase. For these reasons, typically, in Silicon Photonic platform, a ring radius should be >7 μm [18]. This leads to a maximum FSR of ≈1.75 THz.

To overcome this limitation, we rely on a structure based on rings having different radii, to implement a Vernier scheme [15, 16]. However, we propose a non-integer ratios between radii (and hence between rings' FSRs), canceling the periodicity of the whole structure and achieving a FSR-free spectral response.

The numerical design procedure is articulated in three steps, summarized by the flowchart in Fig. 2.4:

1. We start with the design of a Vernier filter with integer FSR ratios, using well established techniques for the synthesis of coupled resonator filters [19–21]. As the target FSR $FSR_t = 4.8$ THz (this is just a starting point, eventually it will be even wider), we consider all the possible configurations of four MRRs, whose FSR is supposed to be an integer fraction of FSR_t. This means that the FSR of the single MRR could assume values of FSR_t/q_j ($j = 1, 2, 3, 4$ is the ordinal subscript identifying a single MRR). For practical implementation in a silicon photonic platform, the radius should not be too short (as discussed before), and, at the same time the footprint should be kept as small as possible. For this reason, each q_j should be between 3 and 6. We choose the combination $q_j = [5; 3; 4; 5]$, which ensures the lowest spurious transmission peaks in the Drop port response (<-30 dB), and the shallowest out-of-band

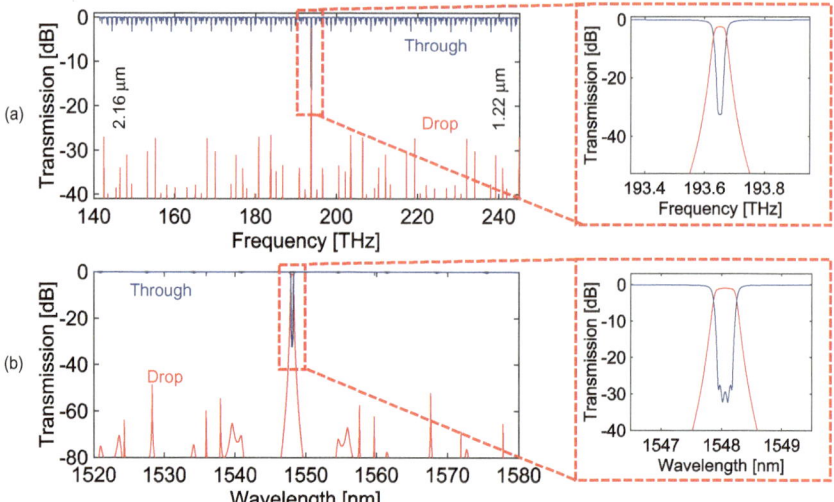

Fig. 2.5 a Simulated spectral responses of the filter (In-Through and In-Drop, respectively in blue and in red) obtained after the second step of the flowchart. **b** Same, after third step. The insets show the spectral shape of the device at resonance frequency/wavelength. Adapted from [22] under creative commons license (creativecommons.org/licenses/by/4.0)

Through port notches (depth <1.5 dB). The power coupling coefficients, at this point, are $K_i = [28.1; 0.8; 0.35; 0.98; 23.6\%]$ ($i = 1, 2, 3, 4, 5$ is the ordinal subscript identifying a single directional coupler). These couplers (i.e., K_i with $i = 1, 2, 3, 4, 5$), providing a Chebyshev response (further optimized in the following steps), are slightly asymmetric since the round-trip losses (approximately 0.02 dB/turn, as discussed in Sect. 2.3.3) have been considered.

The first and last coupling coefficients refer to the point couplers' values (i.e., not TCs implemented by means of Mach-Zehnder Interferometers). This will be further investigated below.

2. We modify the set of q_j (around their integer nominal value) to improve the off-band response of the device, in such a way as to cancel Drop-passband replicas, and, at the same time, keeping low out of band transmission peaks (Drop port) and shallow off-band notches (Through port). After this step we obtain $q_j = [5.73; 3.3; 4.0; 4.8]$ and a single passband in a frequency range larger than 100 THz. Notably, during this step we do not change power coupling coefficients (K_i) and we consider $\frac{dK_i}{d\lambda} = 0$.

   Eventually, the chosen radii are $r_j = [14.6; 8.4; 10.2; 12.2]$ μm, corresponding to $FSR_j = [0.8; 1.4; 1.2; 1.0]$ THz. The simulated spectral responses (Through and Drop) are reported in Fig. 2.5a.

3. We optimize the in-band response of the filter (in the wavelength range of interest, i.e. 1520−1580 nm) by modifying the set of K_i. In this stage we take into account not only

$\frac{dK_i}{d\lambda}$, but also $\frac{dn_{eff}}{d\lambda}$ (waveguide dispersion). The filter response optimization is based on the minimization of the MSE with respect to the target spectral mask, characterized by:

- Drop port 3 dB-bandwidth, $B_{3\,dB} = 40$ GHz,
- Drop port isolation (50 GHz far from central wavelength), $I_{50\,GHz} > 20$ dB,
- Through port return loss (averaged over 20 GHz around the central wavelength), $RL > 18$ dB.

These requirements are related to the optical communication system to which the TOADM is targeted. In particular it has to handle signals whose analog bandwidth is approximately 30 GHz, with a negligible optical cross-talk and quite steep roll-off. The device has to work (mainly) on a 100 GHz-WDM-grid. The return loss constraint, instead, is to avoid coherent interference in the presence of two signals (at the same carrier wavelength), one added and one dropped. We monitor the Bit Error Rate (BER) of the dropped one and we assume the added as interference. Considering the transceiver in [23], to keep the monitored BER below the FEC threshold ($2 \cdot 10^{-2}$), the coherent interference should be attenuated by (at least) 18 dB, thus an $RL > 18$ dB is needed. This value holds for an OSNR ≈ 20 dB.

The optimization occurs at the same time for the filter placed at three different wavelengths corresponding to the beginning, the center and the end of the conventional telecommunication band (C-band), 1520, 1550 and 1580 nm. We obtain coupling coefficients.

$K_i = [28.1; 1; 0.35; 0.85; 23.6\%]$ and noteworthy off-band (Drop port) transmission peaks are always < -30 dB, while (Through port) spurious notches depths are < 1.2 dB. The spectral responses (Through and Drop) of the final device are reported in Fig. 2.5b. The main parameters of the simulated device are reported in Table 2.1, for different wavelengths.

2.3.2 Directional Coupler Design

To effectively realize intra-filter directional couplers (which are both symmetric and asymmetric), Finite Difference Time Domain (FDTD) simulations have been performed. This kind of numerical analysis has been useful not only to optimize the single K_i, but also to quantify the coupler-wavelength sensitivity.

Without going too much into the details and just as an example, we report in Fig. 2.6a and b the sketch and the cross-section (respectively) of the directional coupler between the second and the third ring. The simulation results are reported in Fig. 2.6c, where the power coupling ratio as a function of the gap between the waveguides implementing the two rings is reported, for three different target wavelength channels (1525, 1545 and 1565 nm).

Table 2.1 Performance of the simulated $4th$ order non-integer Vernier filter, accounting waveguide dispersion and coupler-wavelength dependence

Vernier filter (nm)	3 dB bandwidth (GHz)	50 GHz channel isolation (dB)	Max drop port isolation (dB)	Max out of band notch depth (dB)
Specifications	40.0	20.0	18.0	1.2
1520	39.5	25.2	25.1	1.2
1550	43.9	22.5	28.0	1.2
1570	47.9	20.1	20.5	1.2

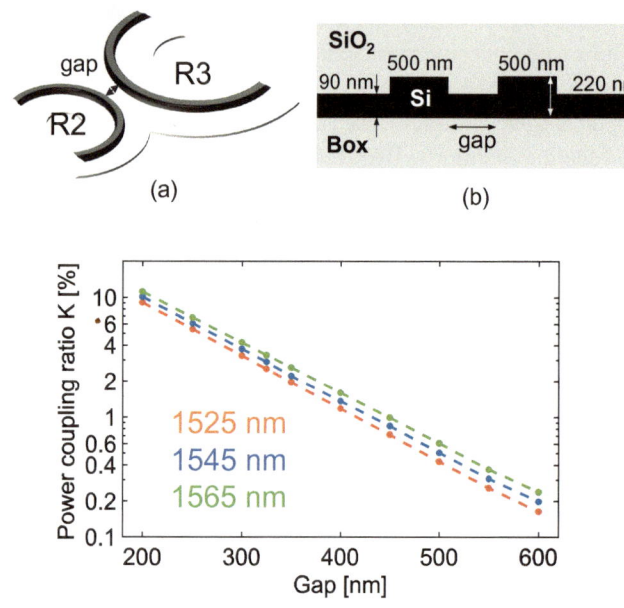

Fig. 2.6 a Sketch of the directional coupler between the second and third ring and **b** its detailed cross-section. **c** FDTD-simulated behaviour of the power coupling ratio with respect to the gap between the two MRRs, for three different wavelengths, specifically 1525 , 1545 and 1565 nm (in orange, blue and green, respectively). Adapted from [22] under creative commons license (creativecommons.org/licenses/by/4.0)

At this point, for convenience, the following parameter is introduced:

$$S(g) = \frac{K_{3,1565}(g) - K_{3,1525}(g)}{K_{3,1545}(g)}. \tag{2.1}$$

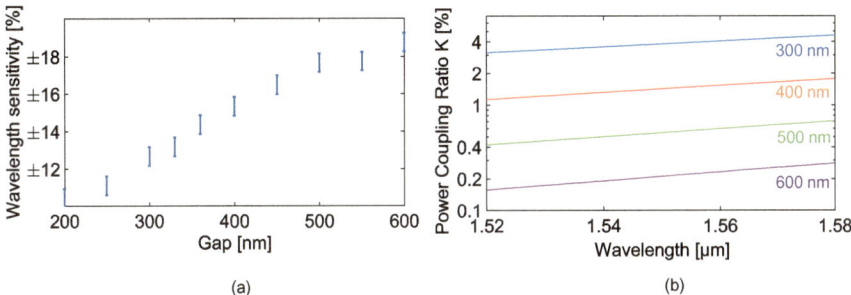

Fig. 2.7 a Wavelength sensitivity (S) evolution with respect to the gap distance. **b** Power coupling ratio against wavelength, for different values of gap distances. Adapted from [22] under creative commons license (creativecommons.org/licenses/by/4.0)

This quantity gives an idea of the wavelength sensitivity of K_3, as a function of the gap (g). In Eq. 2.1, the subscript x in $K_{3,x}$ identifies the wavelength.

In Fig. 2.7a, S is plotted against g[nm] and the highest sensitivity (approximately 0.18) can be observed for the widest gap, while the lowest sensitivity (approximately 0.1) is at the narrowest gap. Finally, in Fig. 2.7b, the behavior of K_3 as a function of wavelength is reported (for different values of gap). Indeed, $\frac{dK_i}{d\lambda}$ must be known and taken under control, since it is one of the sources of performance variability for the whole filter along the operational wavelength. To correctly implement $K_3 = 0.35\%$ we choose a gap of 530 nm.

To mitigate as much as possible the impact of coupling-wavelength dependence on the filter, we use TCs to connect the rings chain with the two bus waveguides [the single TC is implemented by means of a tuneable Mach-Zehnder Interferometer (MZI)]. To obtain the same power coupling ratio of a directional coupler with a MZI, we rely on the following equation, to obtain the same power coupling ratio of a directional coupler with a MZI:

$$K_x = \sin^2(0.5\sin^{-1}(\sqrt{K_{MZIx}})), \tag{2.2}$$

where K_x (with x = 1 or 5) are the point directional couplers, that constitute the (top/bottom, respectively) interferometer itself, and K_{MZIx} is the coupling ratio the MZI has to implement. Figure 2.8a clarifies the relation between K_{MZIx} and K_x and Fig. 2.8b shows the final designed filter. According to Eq. 2.2, with $K_{MZI1} = 28.1\%$ and $K_{MZI5} = 23.6\%$, it results that $K_1 = 7.6\%$ and $K_5 = 6.3\%$.

The outcome of FDTD simulations for K_1 (symmetric coupler, since the bending radius of the first ring and of the waveguide above it are the same and equal to 14.6 μm) are reported in Fig. 2.9a (power coupling ratio against gap, for the three different wavelengths, 1525 nm, 1545 nm and 1565 nm) and b (wavelength sensitivity of power coupling ratio as a function of the gap). In the end, a gap of 250 nm is needed to achieve the desired coupling ratio ($K_1 = 7.6\%$).

Eventually, the set of power coupling ratios is $K_i = [(7.6, 7.6); 1; 0.35; 0.85; (6.3, 6.3\%)]$.

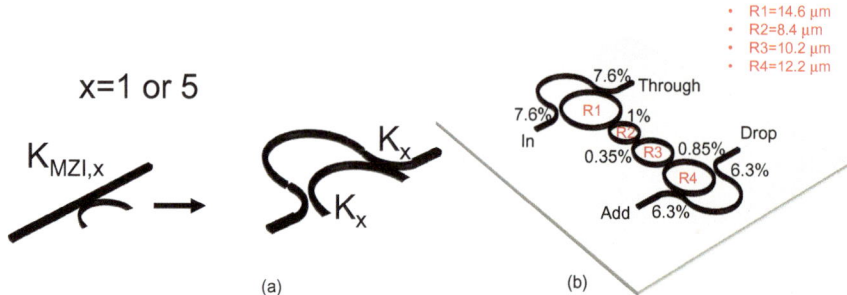

Fig. 2.8 **a** Relationship between typical directional coupler and MZI-based TC. **b** Final designed filter, with details regarding radii and coupling ratios

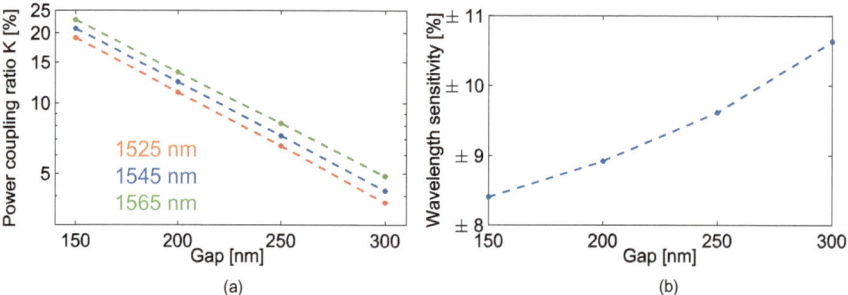

Fig. 2.9 **a** FDTD simulation of power coupling ratio K_1 (between symmetric bending radii, $r = 14.6$ μm) for different values of gap distances and for different wavelengths. **b** Its wavelength sensitivity (S) versus gap distances. Adapted from [22] under creative commons license (creativecommons.org/licenses/by/4.0)

By tuning the upper arm of this MZI (through a thermo-optic actuator) we can make the power coupling ratio between the first/last MRR and the upper/lower bus waveguides almost constant across the whole operative wavelength range (i.e., the best impedance matching is always guaranteed [19, 24]).

The same FDTD simulations and considerations have been carried out for coupler K_5.

In the following table the relevant parameters of the (numerically simulated) filter for the different wavelength channels are summarized and can be compared with the expected specifications, which will be later valid for testing purposes.

2.3.3 Hitless Tuneability

As already discussed in the introduction of this chapter, the possibility of hitlessly tuning (i.e., reconfiguration of the filter from one wavelength channel to another, without impairing the signals whose carrier wavelength is in the between) an Add/Drop multiplexer is very

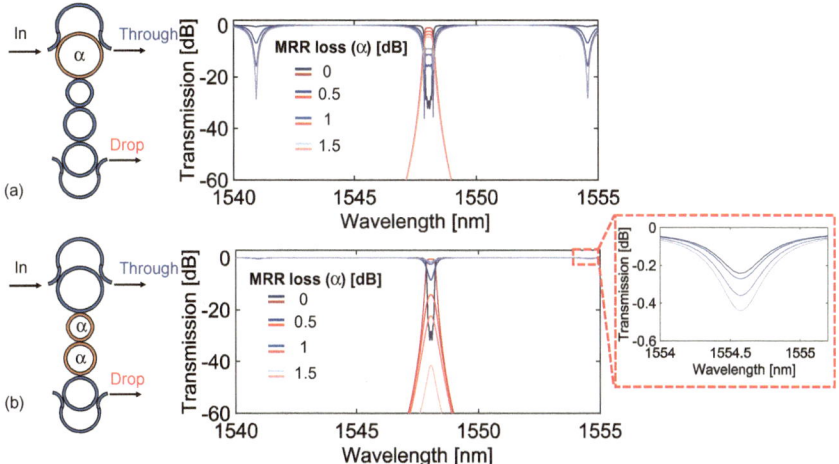

Fig. 2.10 a Simulated spectral responses (In-Through and In-Drop, respectively in blue and red) in the case of extra round-trip losses introduced in the first ring of the chain. Deep spurious notches appear for just 1 dB of extra loss. **b** Same as (**a**), but introducing excess losses in the second and third rings. Spurious notches are much shallower, in this case (<0.6 dB deep). Adapted from [22] under creative commons license (creativecommons.org/licenses/by/4.0)

appealing. In this regard, there are a few successful examples in the literature, that use switches to disconnect the filtering structure from the network buses [17]. However, on the one hand this strategy is not feasible for Vernier-based topology since it may lead to the appearance of off-band notches [25], and on the other hand it implies the necessity of the circuit duplication [26], which is usually not convenient in terms of complexity and area occupation.

In this work, we perform hitless tuning by increasing the round-trip losses (α) of the filter's MRRs. The choice of the rings to be controlled is crucial. In Fig. 2.10a the (simulated) transfer functions of the device (for both Through and Drop ports) are reported, for different values of α ($\alpha = 0, 0.5, 1, 1.5$ dB/turn), applied to the first ring. As it can be observed, as the round-trip losses increase the first MRR approaches the critical coupling condition [27], leading to deep notches (>30 dB), in Through port transmission, without disconnecting the Drop passband (its transmission attenuation is less than 5 dB). The spectral distance between consecutive notches (or between passband and notches) is the first ring's FSR. If α becomes noticeably greater than 2 dB the filter is properly disconnected, but the transient itself (i.e., when the α passes from 0 to 1.5 dB) does not match the definition of "hitless".

For this reason, the exploitation of the attenuation of the second and third rings is more convenient, as also shown in Fig. 2.10b. When the round-trip losses of both MRRs reach 5 dB the filter is properly disconnected from the bus waveguide and the depth of out-of-band notches of the Through port response is always less than 0.6 dB.

Notably, theoretically speaking, increasing the round-trip loss of only one ring should be enough, but it would require an attenuation not practically achievable in such a small footprint.

The exploited mechanism to intentionally induce round-trip losses is, in fact, FCA, induced by free-carrier injection, according to the following empirical relation [28] (valid for silicon approximately at 1550 nm):

$$\Delta\alpha[cm^{-1}] = -\left[8.5\frac{N_e}{cm^{-3}} + 6.0\frac{N_h}{cm^{-3}}\right] \cdot 10^{-18}, \tag{2.3}$$

where N_e and N_h are the electrons and holes concentrations (in cm^{-3}), respectively. The lateral slab is p-doped (10^{20} cm^{-3}) and n-doped (10^{20} cm^{-3}) at a distance (d) of 900 nm from the waveguide core (we define it as "clearance"), around the two innermost MRRs. In so doing a Variable Optical Attenuator (VOA) is actually implemented, and by forward biasing it, the carriers are injected [29–31] in the waveguide core and the cavity-enhanced losses are useful to impede the optical signal transmission. Furthermore, these losses (proportional to the carrier density in the waveguide core) are controllable by adjusting the voltage applied to this p-i-n junction (the waveguide core is indeed intrinsic). Figure 2.11a shows the 3D schematic of a single ring, surrounded by doped regions, while in the inset a detailed waveguide cross-section is reported. Figure 2.11b shows the two inner rings of the fabricated filter, where the VOAs are placed (the n-doped region is in the outer part of the resonators).

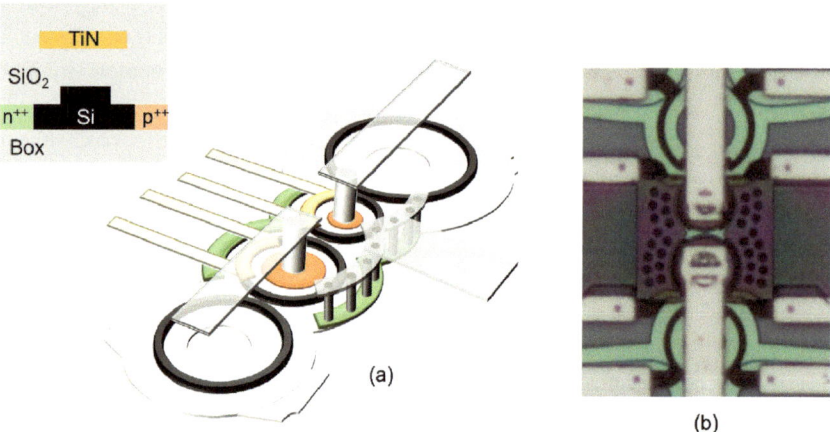

(a)

(b)

Fig. 2.11 **a** 3D schematics of the p-i-n junctions surrounding the second and the third ring of the chain, with detailed cross-section. The p and n-doped regions are highlighted, in orange and green, respectively. **b** Top-view microphotograph of the fabricated MRRs. Adapted from [22] under creative commons license (creativecommons.org/licenses/by/4.0)

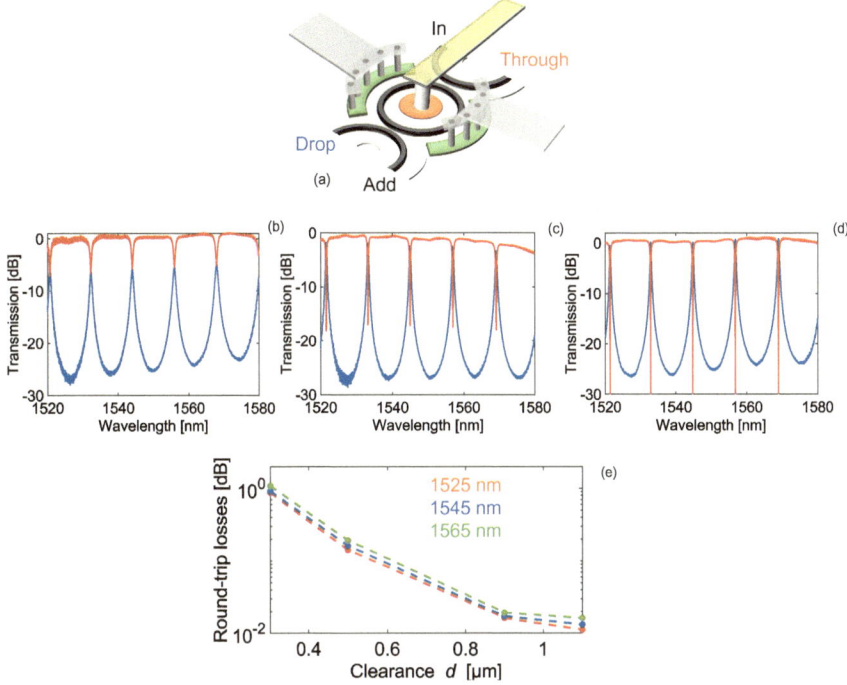

Fig. 2.12 **a** Scheme of an MRR surrounded by the p and n-doped regions. Spectral responses of this structure (Through port, in blue and Drop port, in orange) measured for different values of clearance, $d = 0.3 \ \mu$m (**b**), $d = 0.9 \ \mu$m (**c**), 1.1 μm (**d**). **e** Fitted values of round-trip losses with respect to clearance, for three different wavelengths. Adapted from [22] under creative commons license (creativecommons.org/licenses/by/4.0)

Usually, the doped regions close to a silicon photonic nanowire can cause unwanted extra propagation losses, even when no voltage is applied between them. These extra losses increase when the clearance decreases, and we investigated the impact by measuring [by means of a Tuneable Laser Source (TLS) synchronized with an Optical Spectrum Analyzer (OSA)[2]] the spectral responses of different MRRs [whose scheme is shown in Fig. 2.12a], with different VOA configurations (i.e., different values of d). For these measurements the VOAs are, of course, unbiased. In Fig. 2.12b, c, d the transfer functions of three identical MRRs, with different values of clearance ($d = 0.3, 0.9, 1.1$ μm), are reported. From those spectra, by using a numerical fit, the round-trip losses can be extracted and the results are reported in Fig. 2.12e.

The round-trip losses appear to be approximately 1 dB/turn for $d = 300$ nm, but are negligible for $d = 1100$ nm (approximately 0.015 dB/turn, in line with the losses of an MRR devoid of VOA). The losses are slightly higher for $d = 900$ nm (approximately 0.02

---

[2] TLS is a 4320A model, while OSA is a 6371. Both of them are from ANDO.

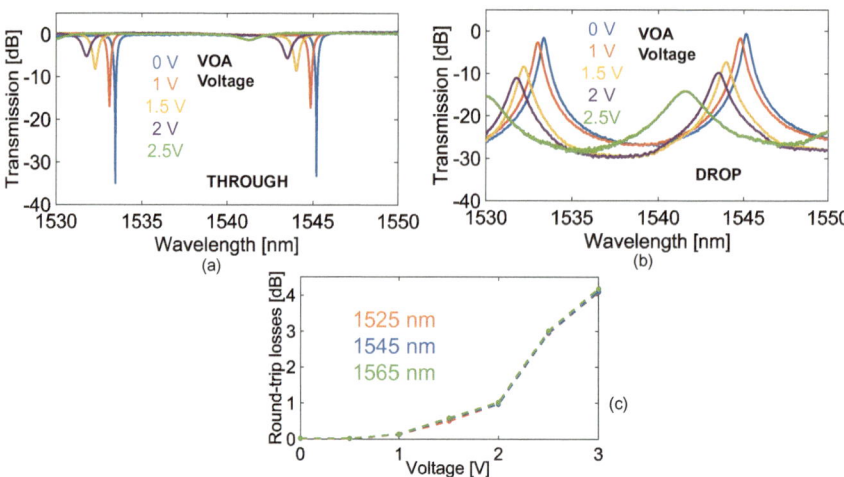

Fig. 2.13 Spectral responses of a single MRR (whose radius is 8.23 μm), In-Through (**a**) and In-Drop (**b**), when there is a voltage drop onto its surrounding p-i-n junction, from 0 V to 2.5 V. **c** By fitting these curves with the theoretical transfer function of the MRR we are able to estimate the round-trip losses as function of the applied voltage. Adapted from [22] under creative commons license (creativecommons.org/licenses/by/4.0)

dB/turn), which is the choice for our design (this configuration also guarantees a reasonable driving voltage to perform a complete hitless operation, as discussed in the next lines).

To evaluate the performance of those p-i-n junctions, we measure the spectral responses (Through and Drop ports) of the MRR, equipped with a VOA whose clearance is 900 nm, by forward biasing the diode with increasing driving voltage (from 0 V to 2.5 V). Figure 2.13a–b show the acquired transfer functions. When the driving voltage increases, the carrier density in the core waveguide increases, and the round-trip loss of the MRR increases. Thus, the Through port main notch becomes less and less deep and the Drop port passband results to be severely attenuated. For example when the driving voltage is equal to 2.5 V, the passband peak is at −14 dB and the depth of Through port notch is 1 dB. By using the same numerical fitting introduced before to assess the round trip losses, we are able to estimate α for this MRR as a function of voltage drop between the p and n-doped regions (Fig. 2.13c). When the applied voltage is approximately 3 V, $\alpha = 4$ dB/turn, suitable for hitless operation.

However, when we inject free-carriers in the waveguide not only FCA occurs, but also Free Carrier Dispersion (FCD). The waveguide n_{eff} changes by the amount Δn_{eff} according to the following empirical equation [28] (again valid for silicon, at 1550 nm):

$$\Delta n_{eff} = - \left[8.8 \cdot 10^{-4} \frac{N_e}{cm^{-3}} + 8.5 \left(\frac{N_h}{cm^{-3}} \right)^{0.8} \right] \cdot 10^{-18}. \qquad (2.4)$$

Moreover the temperature of the waveguide dramatically rises due to the dissipated power of the diode (driven above its own threshold) [32]. As it can be observed from the measurements reported in Fig. 2.13a–b, while the disconnection occurs, the ring resonance evidently shifts toward shorter wavelengths (blue shift). Since the waveguide heating causes a red shift (ring resonance moving toward longer wavelengths), we can conclude that the FCD is dominant in this situation. It must be noted that the FCD and the thermal effects do not have the same dynamics. In fact, the latter has a time constant of tens of microseconds [33], while the former is much faster (approximately 1 ns [34]), compliant with the switching time needed by typical optical networks. Further considerations on the time domain evolution of these phenomena are discussed in Sect. 2.4.

2.3.4 Polarization Transparency

Even though the TOADM presented in Sect. 2.2 can handle just one SOP, it is worth mentioning that we foresee the polarization transparency for the single filter. Polarization independent MRR filters are not new in the literature [12], however this feature has never been combined with hitless operation.

Silicon photonic nanowires and circuits show a behaviour, that strongly depends on the SOP of the light. Thus, the most effective approach to handle the two input orthogonal polarization states is to treat them separately [12]. To achieve the polarization insensitivity we exploit the topology sketched in Fig. 2.14a, which is the cascade of a Polarization Splitter and Rotator (PSR) [10], two nominally identical filters and a Polarization Rotator and Combiner (PRC) [10]. The top-view microphotograph of this device is instead reported in Fig. 2.14b.

The PSR splits the incoming signal (either coming from the In port or from the Add port) into its own orthogonal polarization states (TE and TM), and rotates TM state, in such a way that only TE mode is present in the photonic circuit. Since the two signal paths may not have the same losses, there are two p-i-n VOAs in one of the two arms of the polarization diversity device.[3] Then, there are the two identical filters, tailored to handle only the TE state. Immediately after the filter a PRC (for both Through and Drop ports), rotates the portion of the signal which was not rotated by PSR (so that each mode is rotated just once) and combines the two polarization states again.

To effectively implement this architecture, waveguide crossings are fundamental [12] since, having two inputs (namely In and Add) and two outputs (namely Through and Drop), the cross-intercept of the two signals coming out from the PSR is unavoidable. By measuring specific test structures we can infer that these crossings introduce an excess loss of 0.06 dB

---

[3] As a matter of fact, while balancing the intensity of the two polarizations the VOA induces a negative phase change. However, the exploited transceiver [23] can compensate for a differential group delay (between the polarization states) <15 ps. The refractive index variation is approximately 0.1 (in case of $10^{20} \mathrm{cm}^{-3}$ carrier concentration in the waveguide core) over a length of 250 μm. Thus, this differential phase change does not impact the transmission quality.

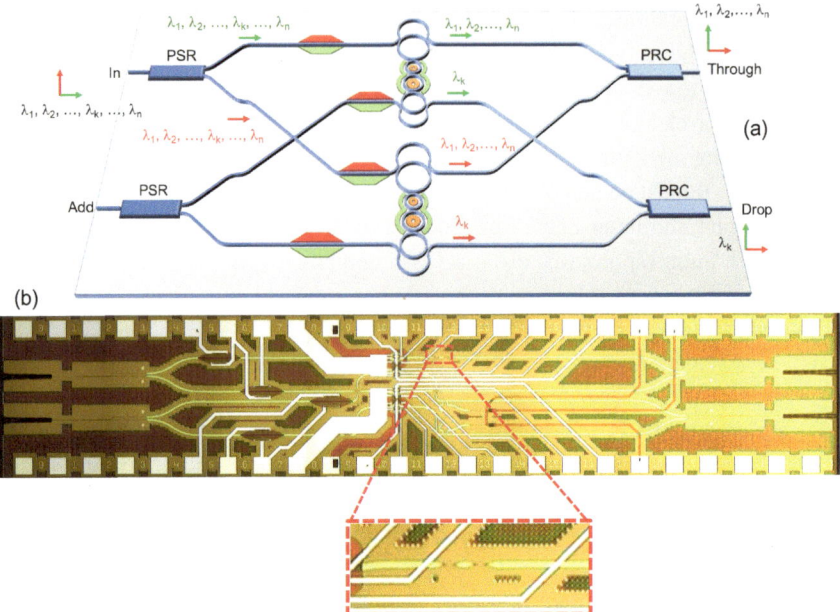

Fig. 2.14 **a** Scheme of the polarization diversity PIC, composed of the cascade of PSR-two nominally identical filters-PRC. **b** Top-view microphotograph of the implemented PIC. PSRs and PRCs are building blocks from [10]. In the inset, a dummy crossing can be recognized, placed to finely equalize optical lengths between branches. Adapted from [22] under creative commons license (creativecommons.org/licenses/by/4.0)

and an optical cross-talk of −40 dB, while there is no evidence of polarization rotation effects. Notably, two dummy crossings are introduced in the two polarization arms [see inset of Fig. 2.14b] to mitigate crossing excess loss on the overall Polarization Dependent Loss (PDL) of the photonic circuit.

To estimate the quality of this polarization diversity scheme, some measurements intended to quantify PDL have been performed. For this purpose we consider a test structure, shown in Fig. 2.15a, composed of a PSR, two straight nanowires 5 mm long and a PRC. By using the same TLS and OSA previously mentioned, we measure the spectral response of this device. The probing signal, before being coupled with the test architecture, has been scrambled in polarization by using a proper polarization scrambler (placed right after the TLS). The test was repeated ten different times. The outcome of these measurements is shown in Fig. 2.15b. The average insertion loss due to the PSR, the PRC and the waveguide propagation is less than 1 dB (nanowire attenuation is supposed to be 1 dB/cm, as further discussed in the next section) across more than 60 nm (1520−1580 nm), while the absolute value of the PDL is below 0.5 dB, which is good enough for the aim of the PIC.

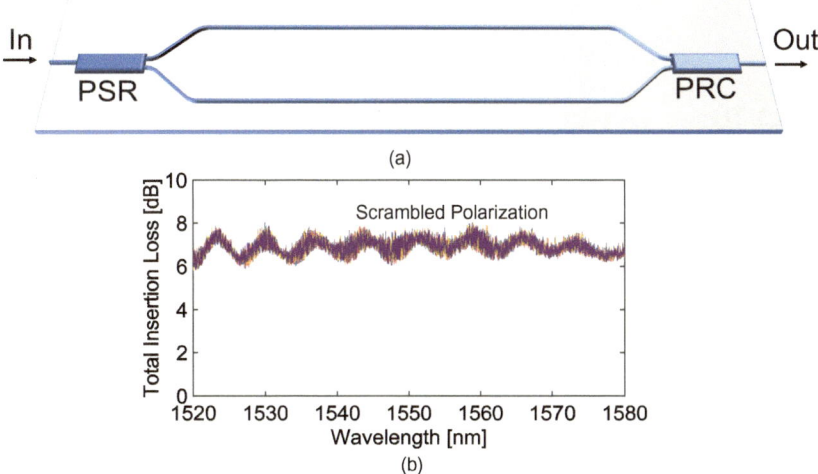

Fig. 2.15 **a** Scheme of the test structure used to estimate the quality of the polarization diversity scheme, composed of the cascade of PSR-two identical straight waveguides-PRC. Waveguides length is 0.5 cm. **b** In-Out spectrum of the test-structure, including I/O coupling loss, obtained scrambling the polarization of the probing signal, ten different times. PDL is <0.5 dB. Adapted from [22] under creative commons license (creativecommons.org/licenses/by/4.0)

2.4   Silicon Photonic Add/Drop Filter Operation

In this section the experimental proofs of the single photonic filter functionalities are provided. Specifically we demonstrate the ultra-wide operational wavelength range, hitless tuning and polarization transparency.

Furthermore, system-level measurements were performed, by using a commercial transceiver, to ensure the behaviour of the device in a realistic environment. Before going into the details, we have to make a premise on fiber-to-chip coupling, which is essential to perform all the following measurements. Both the complete TOADM and the test structures hosting the single filter (single or double polarization) are provided with suitable mode adapters [35] (i.e., suspended tapers from [10], non polarization selective). The best coupling is achieved by using a small core fiber (UHNA7, Nufern), whose Mode Field Diameter (MFD) is 3.2 μm. We perform insertion loss measurements of the silicon nanowires (0.5 cm and 7.1 cm long), by using this equipment, across the wavelength range of 1520−1580 nm (TE-polarization), as shown in Fig. 2.16. We can infer a propagation loss of approximately 1 dB/cm, a Wavelength Dependent Loss (WDL) of 0.2 dB and a coupling loss of 3 dB/facet. This last value might not be compliant with the application of the PIC, but recently solutions to achieve fiber-to-waveguide losses <2 dB, over more than 100 nm have been proposed [36].

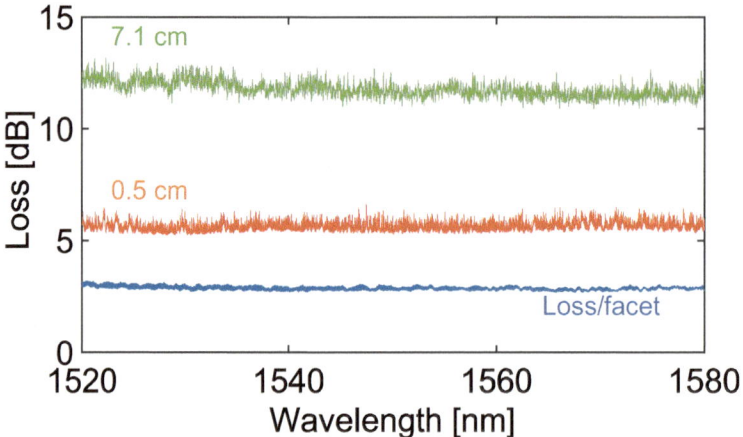

Fig. 2.16 In-Out spectrum of a silicon nanowire 0.5 cm-long (orange curve) and 7.1 cm-long (green curve), equipped with I/O suspended couplers. Facet losses are reported as well (blue curve). Adapted from [22] under creative commons license (creativecommons.org/licenses/by/4.0)

Anyhow, the silicon chip hosting the single filter (single or double polarization) or other test structures is glued and wire-bonded onto a Printed Circuit Board (PCB), capable of providing/reading suitable electrical signals (defined by the user, by means of custom software) from/to the sensors/actuators, as shown in Fig. 2.17. The manual mechanical stages ensure I/O fiber-chip coupling, while the temperature of the chip-substrate is controlled (and held at 25°C) by using a Thermo-Electric Cooler (TEC) controller, which reads the actual temperature through a thermistor (in the vicinity of the PIC) and drives a Peltier cell placed under the chip, according to a Proportional Integral Derivative (PID) control scheme (whose accuracy is ± 0.05 K[4]).

2.4.1   Ultra-Wide Operational Wavelength Range

Figure 2.18a shows the seven measured spectral responses (Through and Drop ports) for the (single channel, single polarization) filter, across a wavelength range of approximately 90 nm, 1520−1610 nm (limited by the available equipment). This proves that the device is truly aperiodic, since there is no sign of any passband repetition. Furthermore, the spurious transmission peaks (Drop-port response) and spurious notches (Through port) are very sporadic and always <-33 and <1.2 dB, respectively, as expected from the design. In Fig. 2.18b–f

---

[4] Considering (frequency selective) silicon PICs, this uncertainty in temperature stability is directly turned into a wavelength shift of the device spectral response, of around ± 0.5 GHz (temperature— wavelength shift dependence will be further discussed in Chap. 3). Since the passband of the device is 40 GHz, and the target signals for this filter have a (analog) bandwidth not larger than 33 GHz, the impact of this issue is minimal.

Fig. 2.17 Electronic PCB, hosting the photonic chip, exploited to drive actuators and read sensors

these responses are reported on a narrow wavelength range, to better appreciate their spectral features (in terms of B_{3dB}, $I_{50\,GHz}$, RL), which are summarized in Table 2.2.

Typical values at the center of the C-band (1550 nm) are $B_{3dB} = 41$ GHz, $I_{50\,GHz} = 32$ dB and $RL = 20$ dB, compliant with the specifications. Moreover, the Drop-port passband shows an insertion loss $IL = 1$ dB, with ripples <0.5 dB around its central wavelength (Table 2.3).

However, due to the couplers-wavelength dependency, the B_{3dB} passband passes from only 38 GHz to more than 53 GHz. Nevertheless, the Drop-port IL, the $I_{50\,GHz}$ and the Through-port RL are in all the conditions of ≈1 dB, >29 dB and >18 dB, respectively.

At this stage we exploit the custom electronic board to independently drive the TCs and the MRRs phase shifters. As expected, the out-of-fab filters do not show the desired transfer functions (we further discuss this point in the next sections), mostly due to the fabrication tolerances and imperfections. Hence, a tuning technique is needed. During the characterization stage, the throughput and the tuning speed are not constraints, while we must be accurate in the validation of all the aspects of the aforementioned design. For this reason, the figure of merit, according to which the working points of the phase shifters are chosen, is the MSE between the desired frequency mask (see Sect. 2.3.1) and the actual response. The flowchart of the steps to follow, and the conceptual block scheme to execute them, are reported in Fig. 2.19a and b, respectively. It involves the heaters' working points modification, the spectral measurements (by means of the pair TLS and OSA), and the MSE computation. If this quantity is minimized then tuning is complete, otherwise the process has to be repeated. Of course, this approach is time and resource consuming, but also extremely precise.

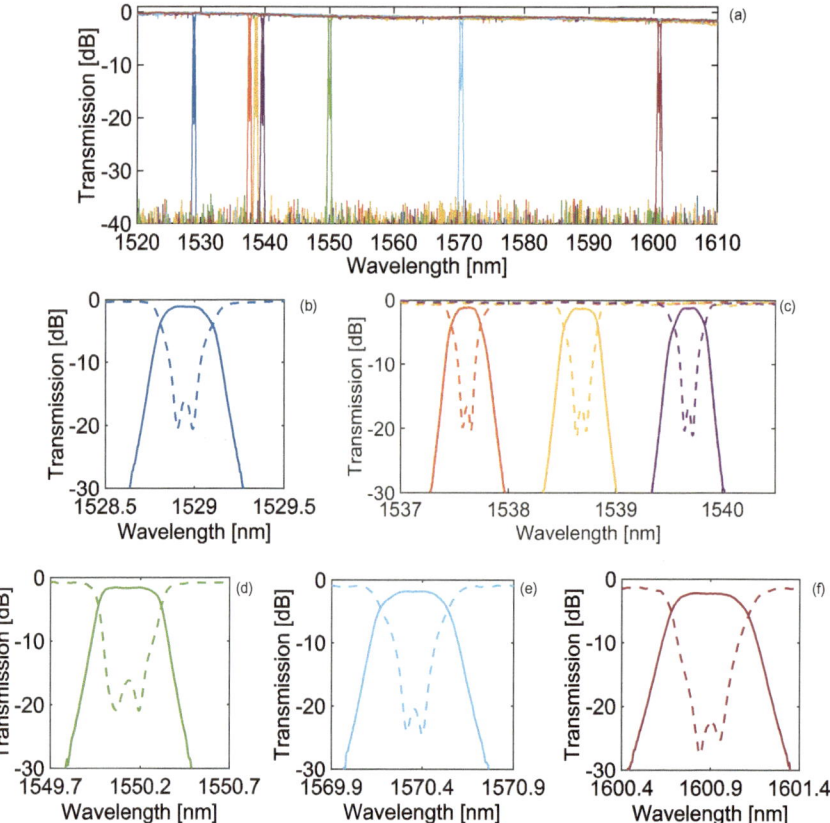

Fig. 2.18 a In-Through and In-Drop aperiodic spectral responses of the single polarization filter, along the C+L band. Specifically, the resonances are placed at 1528.9 nm (**b**), 1537.7 nm, 1538.7 nm, 1539.7 nm (**c**), 1550.2 nm (**d**), 1570.4 nm (**e**), 1600.9 nm (**f**). Performances in terms of bandwidth, return loss or isolation at the distance of 50 GHz from the resonance strongly depend on wavelength. Adapted from [22] under creative commons license (creativecommons.org/licenses/by/4.0)

2.4.2 Hitless Tuning

In Sect. 2.3.2 we discuss the possibility of achieving hitless tuning, by injecting carriers in the waveguide core, here we experimentally demonstrate it (again on a single channel, single polarization filter), through the measurements reported in Fig. 2.20. At the initial stage, the VOAs are unbiased (no voltage applied) and the passband of the filter is at 1529 nm. Then, progressively, the voltage applied to both p-i-n junctions increases, up to 1.3 V (corresponding to an extra round-trip loss of 0.5 dB/turn). At this point we can observe an isolation >35 dB of the Drop port response and a ripple of ≈0.3 dB at the Through port. In

Table 2.2 Performance of the simulated 4th order filter, based on non-integer Vernier scheme

Vernier filter (nm)	3 dB bandwidth (GHz)	50 GHz channel isolation (dB)	Return loss (dB)
Specification	40.0	20.0	18.0
1520 nm	39.5	25.2	25.0
1545 nm	43.9	22.5	28.0
1570 nm	47.9	20.1	20.5

Table 2.3 Measured values of B_{3dB}, $I_{50\,GHz}$, RL and IL for the seven different wavelength channels

Vernier filter (nm)	3 dB bandwidth (GHz)	50 GHz channel isolation (dB)	Return loss (dB)	Insertion loss (dB)
1528.9	38.1	35.0	18.8	1.1
1537.7	39.4	34.7	19.0	1.2
1538.7	39.9	33.9	19.4	1.1
1539.7	40.0	33.2	19.7	1.2
1550.2	41.2	31.9	20.0	1.1
1570.4	43.9	30.1	22.3	1.2
1600.9	52.9	28.9	24.7	1.2

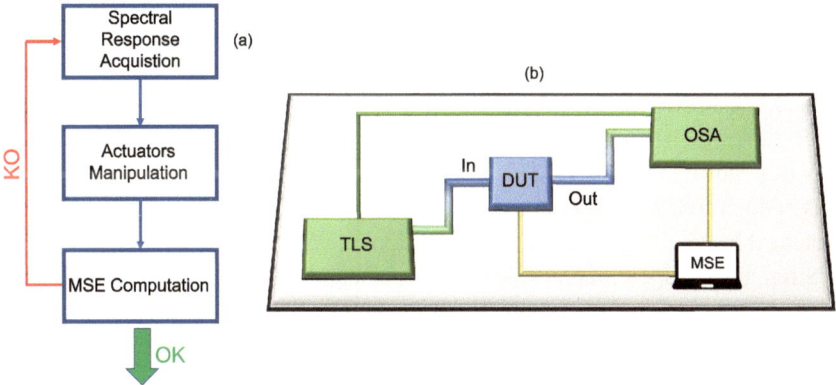

Fig. 2.19 **a** Classical approach to tune a filter with several degrees of freedom. **b** Block scheme of a possible setup to execute these steps. Adapted from [6], with permission

the left inset of Fig. 2.20, the spectral responses for an applied voltage equal to 0, 0.9 and 1.3 V are shown.

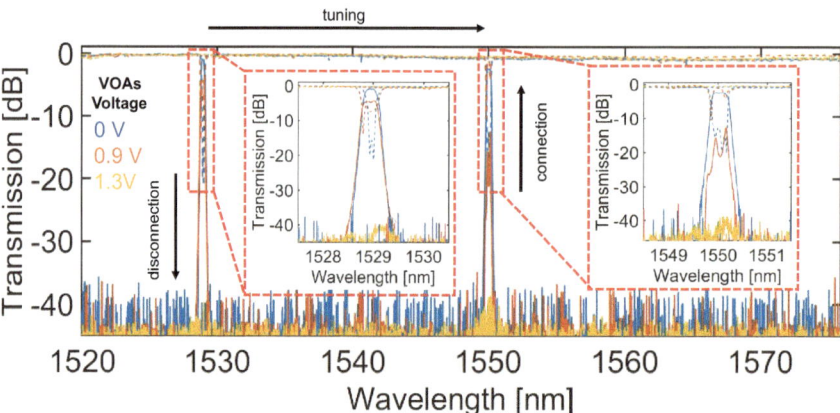

Fig. 2.20 Hitless tuning of the single polarization filter. First, the passband is placed at 1529 nm. For different values of the VOAs driving voltages (0, 0.9, 1.3 V-filter disconnection from the bus waveguide), the spectral responses of the filter, at Through (dashed lines) and Drop (solid lines) ports, are shown in blue, orange and yellow respectively. The same is true for filter re-connection, at 1550 nm. Adapted from [22] under creative commons license (creativecommons.org/licenses/by/4.0)

Notably, to achieve such high isolation, the round-trip losses are not required to be >4 dB/turn (driving voltage >3 V) as stated in Sect. 2.3.2. In the complete device, other phenomena occur, aiding the complete disconnection of the filter. First, the resonances of the two inner rings shifts toward shorter wavelengths (blue shift due to FCD), and all the rings experience a heating (VOA heat dissipation), which should lead to a shift toward longer wavelengths (thermo-optic effect red shift). For the central rings blue shift dominates (see Fig. 2.13). Moving the four resonances to different spectral directions enhances the disconnection, which completely occurs in around 80 ns (and for an applied driving voltage of 1.3 V), as shown in Fig. 2.21a. The blue shift is orders of magnitude faster than red shift, but it is enough to achieve a proper disconnection, in this time frame. This measurement was obtained by coupling a narrow-linewidth signal (whose carrier wavelength was placed at 1528 nm, matched with the position of the passband) to the input port of the filter. By reading the PD response placed at the Drop output of the PIC, while turning on both VOAs we are able to estimate the disconnection time.

After the complete disconnection of the filter, the actuators can be driven to reach the final condition (i.e., filter passband at 1550 nm), which has been properly obtained in the previous section. Now, the filter is ready to be reconnected to the bus waveguide, however this is quite critical, since again the combination of the FCD and the thermal phenomena plays a key role.

These side effects can be effectively counteracted by using thermo-optic actuators. After "switching-off" the filter, all the heaters are driven by adding a "cooling down" voltage (i.e., decreasing their driving voltage), to compensate for the thermal field induced by the VOAs.

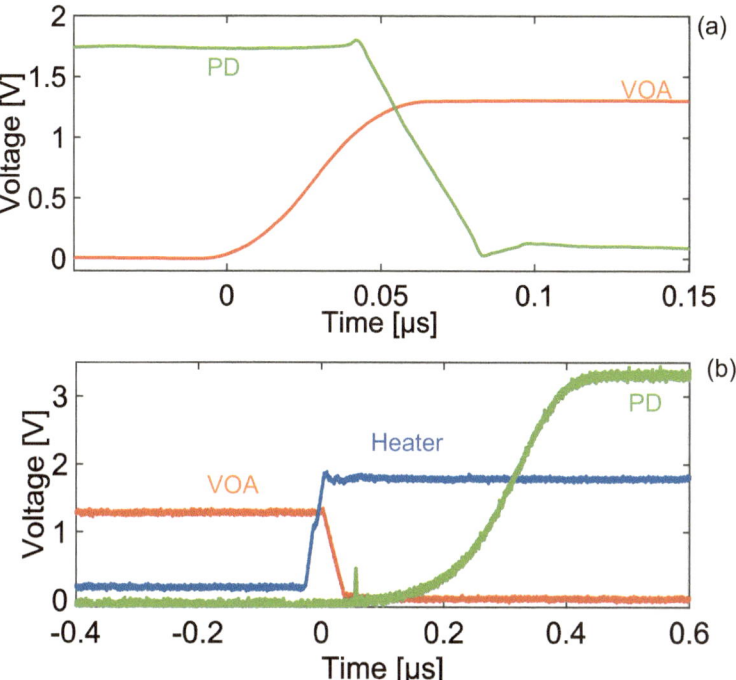

Fig. 2.21 Time domain signals. The orange one is the driving voltage of the VOAs. The green one is the power sensed by the PD, placed at the Drop port of the single polarization filter. The blue one is the (single) heater driving signal. The evolution of the signals is reported during **a** disconnection of the PIC (which occurs in ≈80 ns) and **b** connection (in ≈400 ns), with heater(s) pre-compensation. Adapted from [22] under creative commons license (creativecommons.org/licenses/by/4.0)

This heater "detuning" does not introduce any impairment in the spectral response of the device since it is now disconnected from the bus. When we want the filter to be connected again, we unbias the p-i-n junctions and simultaneously remove the intentional bias from the heaters. As shown in Fig. 2.21b, obtained with the same setup of Fig. 2.21a (but using a probing signal at 1550 nm), the full reconnection occurs in 400 ns (much faster than the thermo-optic effect time constant). This number can be further improved by synchronizing the "cooling down" voltage signals with VOAs driving signals and by optimizing their time-domain shape [37].

As already mentioned in the previous section, during this characterization phase, we use a custom electronic board to drive the thermo-optic actuators and the p-i-n junctions. For a single filter, single polarization, eight actuators (the MRRs' phase shiters, the TCs' phase shifters and the VOAs) should be simultaneously (and independently) turned on. In the worst case scenario, the heaters above MRRs provide a 2π shift, the heaters controlling TCs provide a π shift, and the p-i-n junctions are 1.3 V forward biased (for the complete disconnection of the device). In this condition we measure the power consumption of these

Table 2.4 The worst case scenario power consumption per single embedded element (2π phase shift on each MRR, π phase shift on each TC and the full VOAs disconnection)

Building block	Dissipated power [mW]
Top MZI	18.0
Ring 1	36.1
Ring 2	60.0
Ring 3	51.0
Ring 4	43.0
Bottom MZI	21.5
VOA Ring #2	22.5
VOA Ring #3	22.5

embedded electrical devices, and the results are reported in Table 2.4 (whose content will be useful in the next chapter).

The peak power dissipation is approximately ≈ 230 mW for thermal tuning, and 45 mW for disconnection. However, in hitless operation, the energy (rather than the power) consumption is meaningful, which is <25 µJ, since VOAs are biased for just hundreds of µs. In real optical networks, typical components to accomplish reconfigurable add-drop multiplexing tasks, are the so-called "Nodes-on-a-Blade" (composed of two or more Reconfigurable Wavelength Selective Switches), whose power consumption is in the units of W and reconfiguration occurs in milleseconds or seconds (see for example flex spectrum ROADM NCS2000 from Cisco or True Flex ContentionLess Twin 8 × 24 from Lumentum).

2.4.3 Polarization Transparency

This section is devoted to the analysis of the polarization insensitivity, which is demonstrated using a setup similar to that introduced in Sect. 2.3.4 (exploited to characterize the PSR-PRC test structures), composed of a TLS, a Polarization Scrambler (PS) and an OSA. The block scheme of the experimental setup is provided in Fig. 2.22. The optical probing signal is generated by the TLS, which is coupled to the PS. It generates, at its output, random uncorrelated polarization states, at 1.1 kHz rate (time scale). The scrambled optical wave is coupled to the double-polarization PIC [shown in Fig. 2.14a–b] and its spectral responses are acquired. To do so, TLS sweeps at a speed of 6pm/s. Thus, the SOPs spectral data points, reported in Fig. 2.23, are uncorrelated.

Specifically, these figures show ten different acquisitions of the filter spectral response, acquired with a wavelength sampling step of 1 pm (every point of every function corresponds to a random SOP), with both filters tuned to the same target wavelength (1537.8 nm). In this scenario, the overall PDL is approximately 1.6 dB and, considering that the experimental

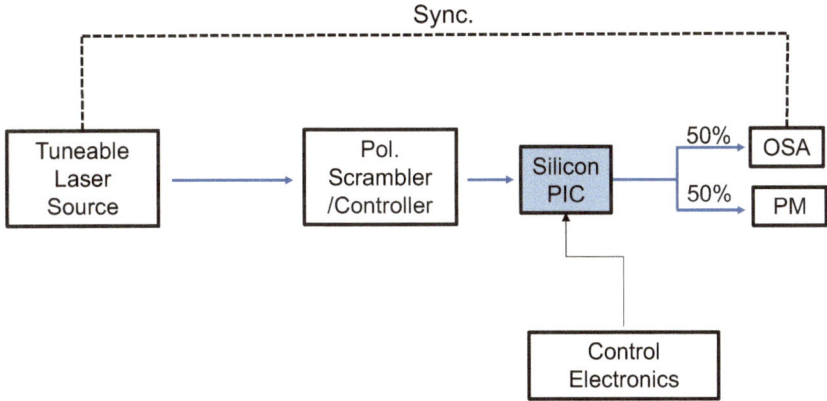

Fig. 2.22 Block scheme of the setup exploited for polarization diversity measurements. TLS and OSA are synchronized, and the probing optical signal before reaching the PIC is scrambled in polarization. The integral output power is read by a Power Meter (PM) and the DUT is equipped with its own control electronics for tuning purposes. Adapted from [22] under creative commons license (creativecommons.org/licenses/by/4.0)

setup surrounding the PIC introduces a PDL of 0.4 dB, we can conclude that the device PDL is ≈1.2 dB. Other spectral features, such as B_{3dB}, $I_{50\,GHz}$ and RL, follow the specification of the device within this wavelength range (i.e., ≈40 GHz, >30 dB, >17 dB, respectively[5]). The wavelength acquisition window spans from 1535 to 1539 nm, and in this spectral region there is no evidence of any spurious notches (Through response) nor of spurious transmission peaks (Drop response).

Since the Drop-port transfer function is all-poles, the Hilbert transform could be conveniently exploited to compute the phase response (net of an offset, related to the device optical length) of the filter at this output [38]. The passband phase against wavelength ($\phi(\lambda)$) is shown in Fig. 2.24a, for all the ten different amplitude responses. The differentiation of $\phi(\lambda)$ over frequency leads to the group delay spectrum ($\tau_g(\lambda)$) and the differentiation of $\tau_g(\lambda)$ over wavelength leads to CD.

The analytical relationships are provided:

$$\tau_g(\lambda) = \frac{\lambda_0^2}{2\pi c}\frac{\partial\phi(\lambda)}{\partial\lambda}, \tag{2.5}$$

---

[5] The typical value for $I_{50\,GHz}$ is 10 dB higher than expected, which is advantageous for the system performance of the TOADM. RL, instead, is not fully compliant with the specifications given at the beginning of this chapter. However, from a practical point of view, this is not a problem. RL should be >18 dB to avoid transmission degradation (in terms of BER), due to coherent interference between two signals coexisting in the same filter (one added and one dropped), when the OSNR of the dropped signal is approximately 20 dB. This minor issue (difference is less than 1 dB) could be easily circumvented by improving the OSNR (2–3 dB is enough) or by increasing the FEC threshold of the transceiver (if it is allowed, as in our case).

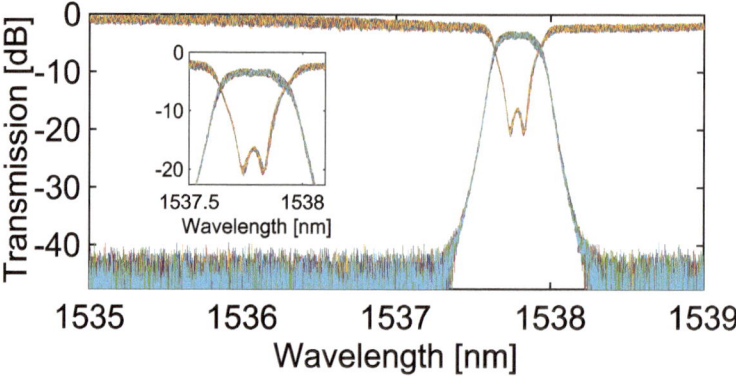

Fig. 2.23 Spectral responses acquired by using the setup in Fig. 2.22 with both filters of the polarization diversity structure tuned. In the inset a zoom on the details close to resonance wavelength. The overall PDL is ≈1.6 dB. Adapted from [22] under creative commons license (creativecommons.org/licenses/by/4.0)

$$CD(\lambda) = \frac{\partial \tau_g(\lambda)}{\partial \lambda}, \tag{2.6}$$

where c denotes the vacuum speed of light and $\lambda_0 = 1538$ nm.

The outcomes of these computations (starting from the ten different amplitude responses) are reported in Fig. 2.24b and c. As further discussed in the next section, the contribution of CD is negligible compared with the compensation margin of the RX stage (more than 10000 ps/nm).

Moreover, the spread of the phase response, the group delay and the dispersion are low, with their standard deviation <0.05 rad, <1 ps and <20 ps/nm, respectively.

Since the filters handling the two polarization states show a similar group delay and the whole structure is balanced from the geometrical point of view, we can state that Polarization Mode Dispersion (PMD) is negligible.

Conversely, the results of the same experiment when only one of the filter is tuned (and the other is at its own out-of-fab status) are quite different, as shown in Fig. 2.25a–d, particularly in terms of PDL. Again ten different amplitude responses (for Through and Drop ports) have been acquired, and the tuned and untuned filters can be easily recognized.

To conclude this dissertation, two points must be addressed.

First of all, even though we foresee the presence of in line VOAs (at the two input waveguides) to equalize the losses experienced by the two polarization states, they have never been driven, during the whole experiment.

Then, it is worth mentioning that the filter shows the very same spectral response (Through and Drop ports) when the input signal is coupled to the two different inputs [see Fig. 2.26a], as in the measurements reported in Fig. 2.26b, with the device tuned approximately 1537.5 nm.

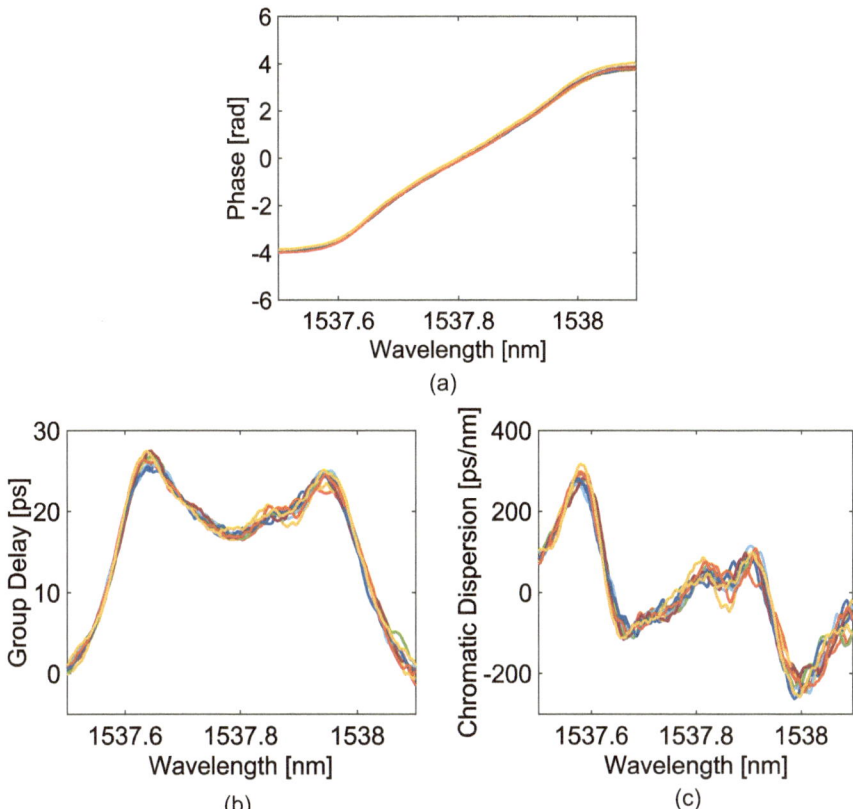

Fig. 2.24 Phase (**a**), Group Delay (**b**) and Chromatic Dispersion (**c**) spectra for the accounted PIC. The phases have been calculated applying the Hilbert transform on the In-Drop amplitude response of each acquisition, reported in Fig. 2.23. The phase spectrum is correct, net on an offset proportional to the PIC optical length. Differentiating phase spectrum, group delay and chromatic dispersion over wavelength are obtained. Adapted from [22] under creative commons license (creativecommons.org/licenses/by/4.0)

2.4.4 System Measurements

To validate the behaviour of this filter in a realistic environment, some system-level measurements are definitely needed. Specifically, we assess the BER in different contexts [we always intend pre-FEC BER], exploiting the setup in Fig. 2.27. We employ a commercial transceiver, namely Jabil Photonics CFP2-DCO [23], capable of generating both 100 Gbit/s Double Polarization-Quadrature Phase Shift Keying (DP-QPSK) and 200 Gbit/s Double Polarization-Quadrature Amplitude Modulation (DP-16QAM) signals, having an output power of −2 dBm, a bandwidth of 32 GHz and a central wavelength tuneable across the C-

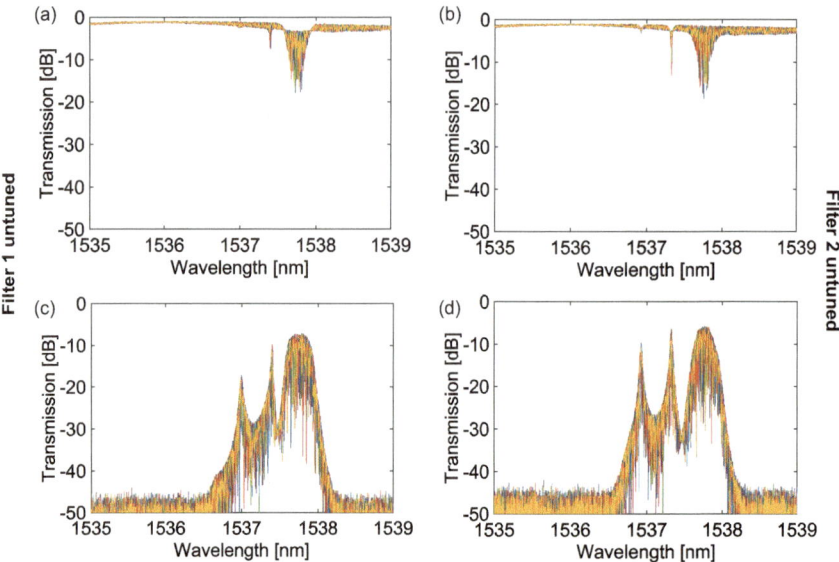

Fig. 2.25 Spectral responses acquired exploiting the setup in Fig. 2.22, with just one of the two filters tuned (the other one is not driven). Again, ten different acquisitions were performed (scrambling the input signal polarization). Adapted from [22] under creative commons license (creativecommons.org/licenses/by/4.0)

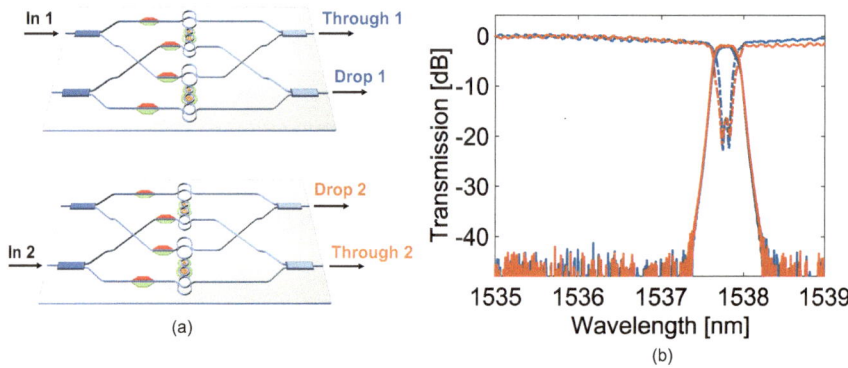

Fig. 2.26 **a** Two possible configurations of the filter. **b** Spectral responses at Through (dashed lines) and Drop (solid lines) ports, when the input signal is coupled to In1 (blue curves) and In2 (orange curves). The spectral overlap is almost perfect. Adapted from [22] under creative commons license (creativecommons.org/licenses/by/4.0)

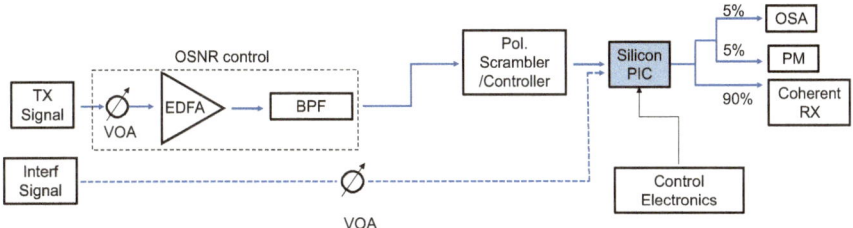

Fig. 2.27 Setup exploited to perform system level measurements. Adapted from [22] under creative commons license (creativecommons.org/licenses/by/4.0)

band (according to the standard 50 GHz-ITU-T grid). The 100 Gbit/s and 200 Gbit/s signal spectra are shown in Fig. 2.28a and b respectively, and overlap with the filter response.

The transmitter (TX) is coupled through an in-line VOA, cascaded with an Erbium Doped Fiber Amplifier (EDFA) (in saturation condition, with an output power approximately 16 dBm) and a 2-nm-bandwidth Band Pass Filter (BPF). These three blocks allow a precise control of the OSNR at the TX side. Immediately after the OSNR control block, the PS can be found. Then, the optical signal is fed to the silicon photonic chip, tuned in a proper condition (passband at 1550 nm). At its output, the light beam is split, by means of 90/10 fiber splitter. 90% of the total power is routed to a coherent receiver (RX), which is equipped with its own Digital Signal Processor (DSP). It can handle input optical power up to 0dBm and has a dynamic range of 18 dB. For all the presented results, the received power was kept constant, approximately −9 dBm.

The RX stage can compensate for:

- CD up to 40000 ps/nm (for 100G signal) or 10000 ps/nm (for 200G signal),
- PDL up to 3 dB,
- PMD up to 15 ps,

and the SOP changes can be tracked up to 300 krad/s. Its Soft Decision Forward Error Correction (SD-FEC) threshold is $2 \cdot 10^{-2}$. The remaining 10% of the optical power collected at the PIC output is coupled to an OSA, in order to evaluate the OSNR. We define it as the ratio of the optical signal power and the noise integrated over a bandwidth of 0.1 nm around the central wavelength (1550 nm).

Figure 2.29a and b show the (pre-FEC) BER as a function of the OSNR, for the 100 G and 200 G transmitted signals, respectively. The black curve represents the back-to-back condition (i.e., when the silicon photonic is substituted by a standard optical fiber), which is almost indistinguishable from the blue curve, that is the case of TX carrier wavelength far away from the passband, propagating in the In-Through bus. This confirms the negligible impact of the fiber-chip I/O couplers and of the pair PSR—PRC.

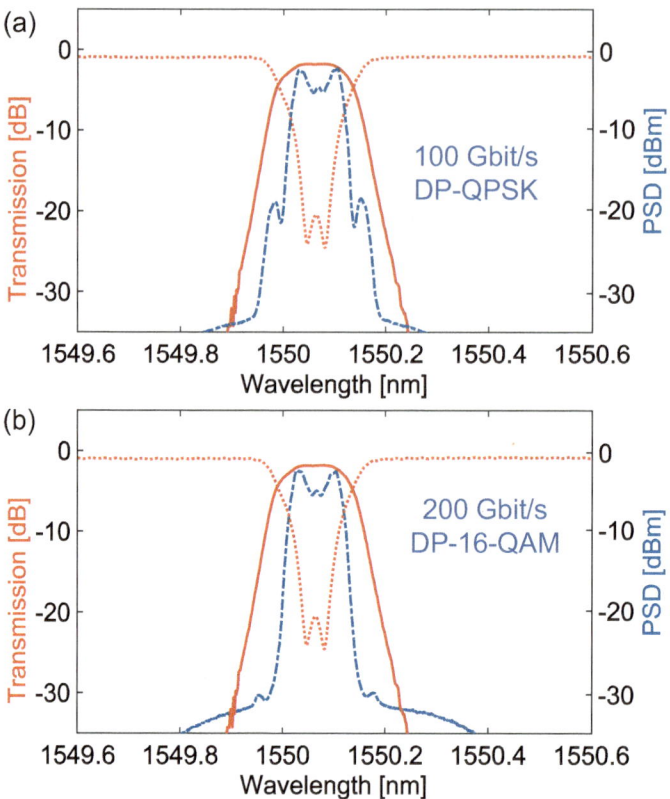

Fig. 2.28 Double Polarization Filter Spectral Response (in orange), for both Through and Drop ports (dashed line and solid line, respectively). PSD (dash-dotted blue line) of the signal exploited to perform system level measurements, **a** 100 Gbit/s DP-QPSK and **b** 200 Gbit/s DP-16QAM

This off-band measurement was repeated changing the carrier wavelength of the TX signal (always far from the passband, namely 1539.77, 1545.32, 1554.12 and 1559.79 nm), to assess the impact of the out-of-band CD (during switching of the filter, i.e., when out-of-band-notches depth is 0.6 dB). Through suitable numerical simulations this CD, caused by the first ring is <40 ps/nm [see Fig. 2.30a], orders of magnitude lower than the maximum CD the transceiver can compensate. Thus, as expected, the BER curves against OSNR for the four mentioned wavelengths show no penalties [see Fig. 2.30b].

Considering again Fig. 2.29a and b, the BER curve referred to In-Drop condition (orange) shows a negligible (or perhaps beneficial, due to noise filtering) filter impact.

At this point, another signal generated by another transceiver is considered (interfering signal). It is a 100 Gbit/s DP-QPSK, whose bandwidth is approximately 28 GHz, with adjustable power in the range [−30; 0] dBm, and with a tuneable carrier wavelength across the C-band, following the standard 50 GHz-ITU-T grid. At first, to validate the impact of

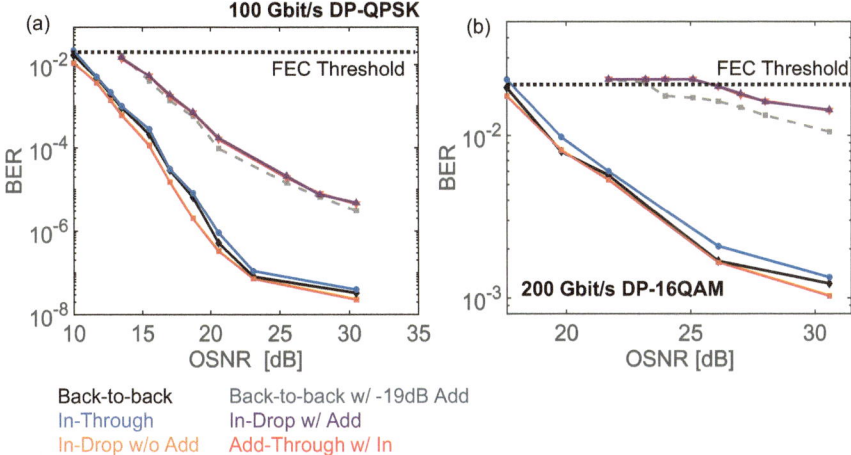

Fig. 2.29 Pre-FEC BER against OSNR for **a** 100 Gbit/s DP-QPSK and **b** 200 Gbit/s DP-16QAM. Adapted from [22] under creative commons license (creativecommons.org/licenses/by/4.0)

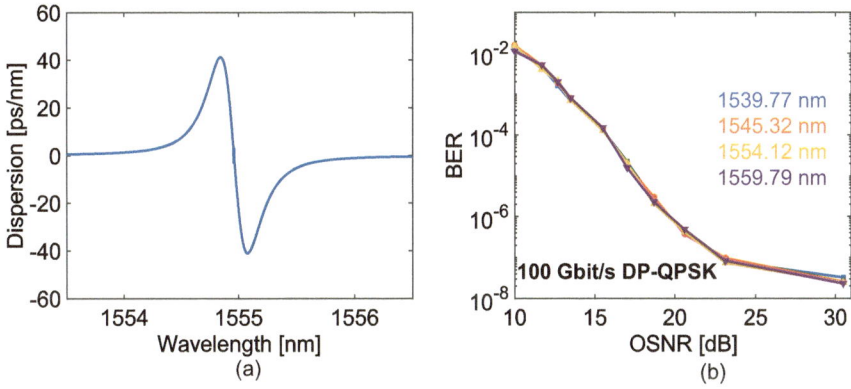

Fig. 2.30 a Simulated Chromatic Dispersion introduced by out-of-band notch, approximated as a first-order MRR filter (with a notch depth of ≈0.5 dB). **b** BER against OSNR curves of the 100 Gbit/s DP-QPSK signal, with the carrier wavelength placed along the out-of-band range of the filter. Adapted from [22] under creative commons license (creativecommons.org/licenses/by/4.0)

the transceiver in presence of an interfering signal, the back-to-back BER measurement has been repeated (grey curve), by coupling the interfering to the back-to-back fiber through a fiber coupler (both signals have the same carrier wavelength, and the interfering signal power is attenuated by 19 dB by using a dedicated in-line VOA, to emulate the typical rejection of the filter).

Then, considering the silicon PIC again, we couple the interfering signal to the Add port of the filter and measure the BER at Drop output. The simultaneous presence of the two

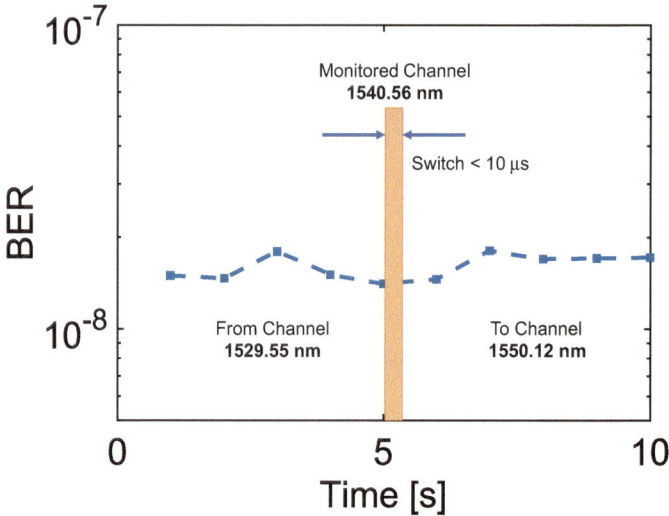

Fig. 2.31 Time domain Pre-FEC BER of the 100 Gbit/s DP-QPSK signal, whose carrier wavelength is at 1540.56 nm, during hitless operation from 1529.55 nm to 1550.12 nm. The BER acquisition rate is 1 Hz. Adapted from [22] under creative commons license (creativecommons.org/licenses/by/4.0)

lightwaves (at the same wavelength) indeed causes coherent interference at the two outputs, which is acceptable for the 100 G situation, but more problematic for the 200G situation (violet curve). The same behaviour (red curve) has been observed by switching the two inputs (i.e., interfering signal coupled to the Input port and the other one coupled to the Add port while measuring the BER at Through output). This states for a symmetrical behaviour of the filter (i.e., $RL \approx 18$ dB), even if its implementation relies on a Vernier scheme.

In general, we can conclude that in almost all the situations (with the exceptions of 200G transmission, with OSNR <25 dB and in presence of strong interfering signal) the BER is below the SD-FEC threshold, which guarantees an error free transmission (since errors are corrected in real time). Furthermore, the same experiments have been carried out by switching off the PS and no significant difference could be observed.

As a final remark, the hitless operation deserves to be investigated by means of system level measurements. In fact, during filter disconnection and tuning spurious notches may overlap with adjacent channels (in the wavelength domain), causing an unwanted attenuation (up to 0.5 dB) and introducing detrimental effects on signal transmission due to the chromatic dispersion. To further understand these aspects, we repeated the experiment in Sect. 2.4.3 (i.e., hitless tuning from the channel at 1529.55 nm to that at 1550.12 nm), while real time monitoring of the BER of the 100G signal, whose carrier was placed at 1540.56 nm. Figure 2.31 shows the BER values over a time window of 10 s (BER samples are acquired once per second). 2 s after the beginning of the observation the filter is disconnected, 3 s later the filter is tuned (this process lasts 10 μs), and again 3 s the passband is connected again.

In the whole time frame no changes in BER can be observed, being always approximately $2 \cdot 10^{-8}$. This means that chromatic dispersion changes are either negligible or within the compensation margin of the RX stage, not impairing transmission.

2.5   Impact of Fabrication Imperfections

After going through the complete design and the full characterization of the considered device, we can make some considerations on possible fabrication imperfections (arising from lithography, etching or wet chemical procedures) and their impact on PIC's system-level performance.

Since we are dealing with interferometric structures, made in a high-index contrast platform (Silicon Photonics), the waveguide shape and dimension are quite critical, in terms of the frequency response of the filter [39, 40].

In the literature, the effects of these fabrication tolerances on MRRs [41] and MZIs [42] have been deeply investigated. In particular, for ring resonators (around 1550 nm), the typical resonance sensitivity on the waveguide width deviations (δw) is $\frac{\partial \lambda}{\partial w} = 1$ nm/nm, while on waveguide height deviations (δh) is $\frac{\partial \lambda}{\partial h} = 2$ nm/nm. Additionally, directional couplers are sensitive to the fabrication variations. In particular, their dependence on δw is $\frac{\partial K}{\partial w} = 10^{-3}$ %/nm, on slab height deviations (δh_s) is $\frac{\partial K}{\partial h_s} = 10^{-2}$ %/nm and on waveguide sidewall angle deviations ($\delta \theta$) is $\frac{\partial K}{\partial \theta} = 10^{-2}$ %/$\circ$ [43].

Since the fabrication imperfections follow, by definition, random distributions, it is quite clear that a complex coupled MRR filter, such as the one described, presents a random "natural" spectral response.

In these regards, some numerical simulations have been performed, including both the MRRs' phase randomness (uncorrelated and uniformly distributed over the range $[-\pi, \pi]$) and the coupling ratios deviations (Gaussian distributed, centered around their nominal value with a $\sigma = 10^{-2}\%$).[6] The results are shown in Fig. 2.32. As it can be observed in most of the cases RL, B_{3dB} and other relevant spectral parameters do not match the specifications reported in Table 2.1.

From these considerations, the necessity of active control[7] arises [46]. Furthermore, optical testing (i.e., checking if the features of a device are compliant with specifications) is almost impossible on a photonic DUT "as-it-is" and a (at least coarse) calibration is definitely needed.

However, as already mentioned in Sect. 2.3, with the actuators embedded in the presented filter, we are able to finely (and individually) tune only the phases of the MRRs, while the impact on other parameters, such as power coupling ratios, cannot be mitigated.

---

[6] Note that these simulations represent the worst case scenario, since usually coherence lengths are few millimeters in Silicon Photonic platforms [44].

[7] Other motivations lead to the implementation of control schemes for PIC, such as the need of fully programmable and reconfigurable photonic devices [45], upon user requirements.

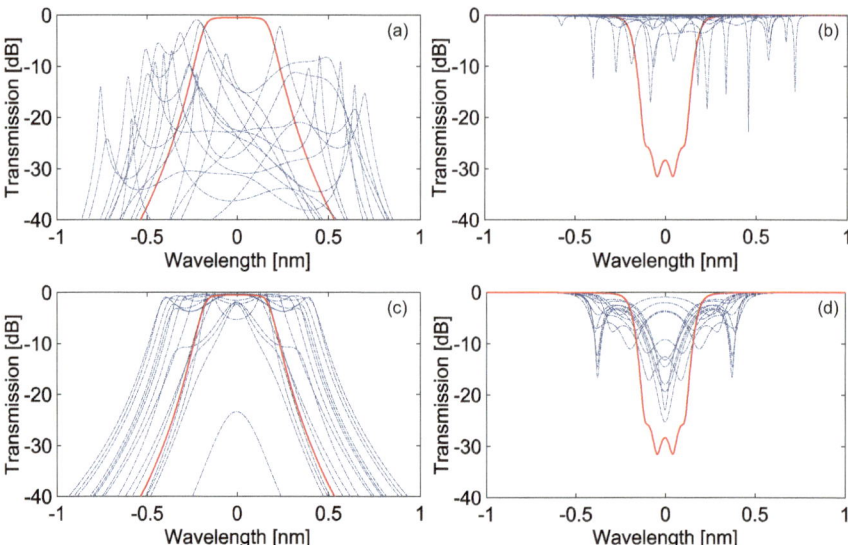

Fig. 2.32 In-Drop (**a**) and In-Through (**b**) simulated spectral responses (dashed blue curves) of the presented filter, with phases intentionally perturbed ($[-\pi, +\pi]$, uniform distribution). In-Drop (**c**) and In-Through (**d**) simulated spectral responses (dashed blue curves) of the presented filter, with the inner directional couplers intentionally perturbed (Gaussian distribution, with $\sigma = 10^{-2}\%$). The solid red lines represent nominal spectral responses

One of the testing aims is, in fact, to state how robust the DUT is to these variations, and how they influence the correct operation of the device. For example, in the case of K_i mismatch, the filter rejection may be <18 dB, having a detrimental effect on the BER performance with a coherent interfering signal.

Regardless, in a volume production environment, a complete characterization (as presented in Sect. 2.4) would be too time and resource consuming, to be performed for every single PIC in a wafer.

Thus, we must translate from a lab characterization to proper testing -which should not be merely a "faster characterization"—to appropriately check and classify PICs, minimizing the errors (i.e., discarding specs-matching devices or selecting non-specs-matching ones). In particular, since the couplers cannot be adjusted, from now on we will focus on the interferometers' phases.

References

1. Q. Cheng, M. Bahadori, M. Glick, S. Rumley, and K. Bergman, "Recent advances in optical technologies for data centers: a review," *Optica*, vol. 5, no. 11, 2018.

2. S. Gringeri, B. Basch, V. Shukla, R. Egorov, and T. J. Xia, "Flexible architectures for optical transport nodes and networks," *IEEE* Communications Magazine, vol. 48, no. 7, 2010.
3. A. Iocco, H. Limberger, R. Salathe, L. Everall, K. Chisholm, J. Williams, and I. Bennion, "Bragg grating fast tunable filter for wavelength division multiplexing," *Journal of Lightwave Technology*, vol. 17, no. 7, 1999.
4. D. Sadot and E. Boimovich, "Tunable optical filters for dense WDM networks," *IEEE* Communications Magazine, vol. 36, no. 12, 1998.
5. D. Smith, J. Baran, J. Johnson, and K.-W. Cheung, "Integrated-optic acoustically-tunable filters for WDM networks," *IEEE* Journal on Selected Areas in Communications, vol. 8, no. 6, 1990.
6. M. Petrini, M. Seyfried, F. Morichetti, and A. Melloni, "Spectral Classification and Cloning of Photonic Integrated Filters for Volume Testing," *Journal of Lightwave Technology*, vol. 41, no. 1, 2023.
7. T. Barwicz, H. Byun, F. Gan, C. W. Holzwarth, M. A. Popovic, P. T. Rakich, M. R. Watts, E. P. Ippen, F. X. Kaertner, H. I. Smith, J. S. Orcutt, R. J. Ram, V. Stojanovic, O. O. Olubuyide, J. L. Hoyt, S. Spector, M. Geis, M. Grein, T. Lyszczarz, and J. U. Yoon, "Silicon photonics for compact, energy-efficient interconnects (Invited)," *Journal of Optical Networking*, vol. 6, no. 1, 2007.
8. E. Klein, D. Geuzebroek, H. Kelderman, G. Sengo, N. Baker, and A. Driessen, "Reconfigurable optical add-drop multiplexer using microring resonators," *IEEE* Photonics Technology Letters, vol. 17, no. 11, 2005.
9. "Nazca Design, Bright Photonics BV, [online]. available: www.nazca-design.org."
10. "PDK of AMF, Advanced Micro Foundry, Singapore, [online]. available: https://www.advmf. com."
11. J. Hryniewicz, P. Absil, B. Little, R. Wilson, and P.-T. Ho, "Higher order filter response in coupled microring resonators," *IEEE* Photonics Technology Letters, vol. 12, no. 3, 2000.
12. T. Barwicz, M. R. Watts, M. A. Popović, P. T. Rakich, L. Socci, F. X. Kaertner, E. P. Ippen, and H. I. Smith, "Polarization-transparent microphotonic devices in the strong confinement limit," *Nature Photonics*, vol. 1, no. 1, 2006.
13. P. Dong, N.-N. Feng, D. Feng, W. Qian, H. Liang, D. C. Lee, B. J. Luff, T. Banwell, A. Agarwal, P. Toliver, R. Menendez, T. K. Woodward, and M. Asghari, "GHz-bandwidth optical filters based on high-order silicon ring resonators," *Optics Express*, vol. 18, no. 23, 2010.
14. Y. Ren, D. Perron, F. Aurangozeb, Z. Jiang, M. Hossain, and V. Van, "Silicon Photonic Vernier Cascaded Microring Filter for Broadband Tunability," *IEEE Photonics Technology Letters*, vol. 31, no. 18, 2019.
15. K. Oda, N. Takato, and H. Toba, "A wide-FSR waveguide double-ring resonator for optical FDM transmission systems," *Journal of Lightwave Technology*, vol. 9, no. 6, 1991.
16. G. Griffel, "Vernier effect in asymmetrical ring resonator arrays," *IEEE Photonics Technology Letters*, vol. 12, no. 12, 2000.
17. R. Chatterjee, M. Yu, A. Stein, D.-L. Kwong, L. C. Kimerling, and C. W. Wong, "Demonstration of a hitless bypass switch using nanomechanical perturbation for high-bitrate transparent networks," *Optics Express*, vol. 18, no. 3, 2010.
18. L. Chrostowski and M. Hochberg, *Silicon Photonics Design: From Devices to System*. Cambridge University Press, 2015.
19. A. Melloni and M. Martinelli, "Synthesis of direct-coupled-resonators bandpass filters for WDM systems," *Journal of Lightwave Technology*, vol. 20, no. 2, 2002.
20. Z. Madsen, *Optical Filter Design*. John Wiley & Sons, 1999.
21. A. Melloni and F. Morichetti, *Componenti e Circuiti per le Comunicazioni Ottiche*. 2010.
22. F. Morichetti, M. Milanizadeh, M. Petrini, F. Zanetto, G. Ferrari, D. Aguiar, E. Guglielmi, G. Ferrari, M. Sampietro, and A. Melloni, "Polarization-transparent silicon photonic add-drop multiplexer with wideband hitless tuneability," *Nature Communications*, vol. 12, 2021.

23. "Jabil Photonics, CFP2-DCO, https://www.jabil.com/industries/photonics/photonics-resources. html."
24. F. Xia, M. Rooks, L. Sekaric, and Y. Vlasov, "Ultra-compact high order ring resonator filters using submicron silicon photonic wires for on-chip optical interconnects," *Optics Express*, vol. 15, no. 19, 2007.
25. M. A. Popović, T. Barwicz, F. Gan, M. S. Dahlem, C. W. Holzwarth, P. T. Rakich, H. I. Smith, E. P. Ippen, and F. X. Kärtner, "Transparent wavelength switching of resonant filters," in *Conference on Lasers and Electro-Optics/Quantum Electronics and Laser Science Conference and Photonic Applications Systems Technologies*, Optica Publishing Group, 2007.
26. H. Haus, M. Popovic, and M. Watts, "Broadband hitless bypass switch for integrated photonic circuits," *IEEE* Photonics Technology Letters, vol. 18, no. 10, 2006.
27. A. Yariv, "Universal relations for coupling of optical power between microresonators and dielectric waveguides," *Electronics Letters*, vol. 36, no. 4, 2000.
28. R. Soref and B. Bennett, "Electrooptical effects in silicon," *IEEE* Journal of Quantum Electronics, vol. 23, no. 1, 1987.
29. H. L. R. Lira, S. Manipatruni, and M. Lipson, "Broadband hitless silicon electro-optic switch for on-chip optical networks," *Optics Express*, vol. 17, no. 25, 2009.
30. Y. Vlasov, W. M. J. Green, and F. Xia, "High-throughput silicon nanophotonic wavelength-insensitive switch for on-chip optical networks," *Nature Photonics*, vol. 2, no. 4, 2008.
31. Q. Xu, S. Manipatruni, B. Schmidt, J. Shakya, and M. Lipson, "12.5 Gbit/s carrier-injection-based silicon micro-ring silicon modulators," *Optics Express*, vol. 15, no. 2, 2007.
32. P. Dainesi, A. Kung, M. Chabloz, A. Lagos, P. Fluckiger, A. Ionescu, P. Fazan, M. Declerq, P. Renaud, and P. Robert, "CMOS compatible fully integrated Mach-Zehnder interferometer in SOI technology," *IEEE Photonics Technology Letters*, vol. 12, no. 6, 2000.
33. M. Jacques, A. Samani, E. El-Fiky, D. Patel, Z. Xing, and D. V. Plant, "Optimization of thermo-optic phase-shifter design and mitigation of thermal crosstalk on the SOI platform," *Optics Express*, vol. 27, no. 8, 2019.
34. S. Meister, H. Rhee, A. Al-Saadi, B. A. Franke, S. Kupijai, C. Theiss, L. Zimmermann, B. Tillack, H. H. Richter, H. Tian, D. Stolarek, T. Schneider, U. Woggon, and H. J. Eichler, "Matching p-i-n-junctions and optical modes enables fast and ultra-small silicon modulators," *Optics Express*, vol. 13, no. 21, 2013.
35. Q. Fang, J. Song, X. Luo, X. Tu, L. Jia, M. Yu, and G. Lo, "Low Loss Fiber-to-Waveguide Converter With a 3-D Functional Taper for Silicon Photonics," *IEEE Photonics Technology Letters*, vol. 28, no. 22, 2016.
36. T. Barwicz, A. Janta-Polczynski, S. Takenobu, K. Watanabe, R. Langlois, Y. Taira, K. Suematsu, H. Numata, B. Peng, S. Kamlapurkar, S. Engelmann, P. Fortier, and N. Boyer, "Advances in Interfacing Optical Fibers to Nanophotonic Waveguides Via Mechanically Compliant Polymer Waveguides," *IEEE Journal of Selected Topics in Quantum Electronics*, vol. 26, no. 2, 2020.
37. M. Harjanne, M. Kapulainen, T. Aalto, and P. Heimala, "Sub-μs Switching Time in Silicon-on-Insulator Mach–Zehnder Thermooptic Switch," *IEEE* Photonics Technology Letters, vol. 16, no. 9, 2004.
38. A. Mencozzi, "A necessary and sufficient condition for minimum phase and implications for phase retrieval," *Trans. Inf. Theory*, vol. 13, no. 9, 2014.
39. A. Melloni, R. Costa, G. Cusmai, and F. Morichetti, "The role of index contrast in dielectric optical waveguides," *International Journal of Materials and Product Technology*, vol. 34, no. 4, 2009.
40. C. Cui and Z. Zhang, "Stochastic Collocation With Non-Gaussian Correlated Process Variations: Theory, Algorithms, and Applications," *IEEE* Transactions on Components, Packaging and Manufacturing Technology, vol. 9, no. 7, 2019.

41. P. Sun, R. G. Beausoleil, J. Hulme, T. V. Vaerenbergh, J. Rhim, C. Baudot, F. Boeuf, N. Vulliet, A. Seyedi, and M. Fiorentino, "Statistical Behavioral Models of Silicon Ring Resonators at a Commercial CMOS Foundry," *IEEE Journal of Selected Topics in Quantum Electronics*, vol. 26, no. 2, 2020.
42. T.-H. Yen and Y.-J. Hung, "Fabrication-Tolerant CWDM (de)Multiplexer Based on Cascaded Mach–Zehnder Interferometers on Silicon-on-Insulator," *Journal of Lightwave Technology*, vol. 39, no. 1, 2021.
43. J. C. Mikkelsen, W. D. Sacher, and J. K. S. Poon, "Dimensional variation tolerant silicon-on-insulator directional couplers," *Optics Express*, vol. 22, no. 3, 2014.
44. Z. Lu, J. Jhoja, J. Klein, X. Wang, A. Liu, J. Flueckiger, J. Pond, and L. Chrostowski, "Performance prediction for silicon photonics integrated circuits with layout-dependent correlated manufacturing variability," *Optics Express*, vol. 25, no. 9, 2017.
45. W. Bogaerts, D. Pérez, J. Capmany, D. A. B. Miller, J. Poon, D. Englund, F. Morichetti, and A. Melloni, "Programmable photonic circuits," *Nature*, vol. 586, no. 7828, 2020.
46. D. A. B. Miller, "Perfect optics with imperfect components," *Optica*, vol. 2, no. 8, 2015.

Electrical Testing

3

3.1 Introduction

In this chapter the electronic infrastructure used to perform the electrical testing and (the preliminary calibration for) the optical testing of the PIC are described in detail. The main features of the proposed electronics are as follows:

1. High number of channels.
   As discussed in the previous chapter, the number of the actuators to be driven and the sensors to be read is quite high (24 heaters, 8 VOAs and 4 PDs, for the 4-channels, single polarization TOADM).
2. 10 kHz-Bandwidth.
   The time response of the system is not critical since the heaters' time constant is fairly slow (approximately 10 μs).
3. High accuracy of DACs and ADCs.
   The heaters, in fact, must be carefully driven (ideally with an uncertainty <1 mV) to correctly tune the TOADM.
   Similarly, the electrical signals generated by the sensors must be properly digitized, to read the proper optical power value.

As discussed in the following paragraphs, these electronics are the combination of a commercial controller (NI-PXI), which actually generates and samples the testing signals, and a custom analog PCB. The latter is needed to provide the PIC correct levels of current (while simultaneously reading them), and, at the same time, recover the electrical signal coming out from PDs.

Nevertheless, in a volume testing scenario, bridging the testing unit with the DUT is a challenge. In particular, in Photonics, the nature of DUT coupling is twofold, electrical and

© The Author(s), under exclusive license to Springer Nature Switzerland AG 2025 49
M. Petrini, *Mixed-Signal Generic Testing in Photonic Integration*, Synthesis Lectures
on Digital Circuits & Systems, https://doi.org/10.1007/978-3-031-60811-7_3

Fig. 3.1 Block scheme of the electrical system designed to perform electrical testing and optical calibration

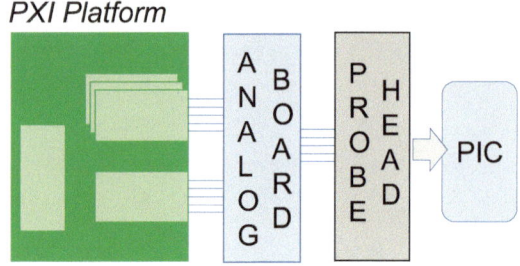

optical. Concerning the former, a suitable PC has been designed, validated and integrated into the system.

The composed system (PXI-custom analog board-PC), shown in the block scheme of Fig. 3.1, is exploited to execute a full (and parallel) electrical test of the DUT, in terms of current-voltage characteristics (I–V curves).

The set of measurements is finally completed by the thermal cross-talk matrix estimation (and its time domain behaviour). The Chapter is organized as follows:

- In Sect. 3.2, the choices of PXI-devices are explained and discussed.
- In Sect. 3.3, the custom analog PCB design is explained.
- In Sect. 3.4, the problem of PIC (electro-optical) coupling access is addressed. In particular, the electrical PC is presented and validated.
- In Sect. 3.5, the electrical measurements (I–V curves) are performed and the relevant parameters of the DUT are extracted.
- In Sect. 3.6, suitable measurements to estimate the cross-talk matrix are presented. Furthermore, its time constant is evaluated.

3.2 PXI Platform

To generate the (electrical) testing signals and to acquire the PIC responses, a system based on a commercial NI-PXI platform has been chosen and designed (its photograph is in Fig. 3.2). The PXI systems are composed of a chassis, which hosts a number of modules (boards) and a controller (that executes the software and acts as an interface between the user and the physical boards).

Such a system can be easily reconfigurable and the boards are interchangeable, making it extremely flexible. Furthermore, the PXI platform could be advantageously exploited in various testing scenarios, offering a high number of channels (for both Input and Output directions), a very tight synchronization among the different modules (through the chassis) and fairly easy programmability with LabView.

Fig. 3.2 Photograph of the PXI system. Some different modules can be recognized

The system we configure is composed of the following:

- PXI controller (PXIe-8881). Having a CPU clock frequency of 4.6 GHz, and capable of handling 32 parallel threads, it can be suitably exploited to test and calibrate the described DUT. Further information can be found in [1].
- PXI chassis (PXIe-1092). It has 9 independent slots, and ensures a synchronization latency <10 ns (with an equivalent jitter of ≈ 1 ns). Further information can be found in [2].
- PXI Analog Output (AO) (PXIe-6738). It has 32 independent output channels (in voltage), ranging in the interval ± 10 V (with a resolution of $300\,\mu$V). The maximum sample update rate is 1 MS/s. The output current is well below 10 mA. Further information can be found in [3].
- PXI Analog Input (AI) (PXIe-6355). It has 80 (single-ended or 40 differential) independent input channels (in voltage), ranging in the interval ± 10 V (with a resolution of $300\,\mu$V).
  The sample rate is 1 MS/s. Further information can be found in [4].

Even if the presented PXI platform could be compliant (in terms of bandwidth and accuracy) with the developed testing routine (which is presented in Chap. 4), it indeed presents a few weaknesses. First and foremost, the PXI AO board is not capable of providing more than a few mA (per channel) to the load. Thus, as it is, it results to be hardly feasible to drive the heaters and the VOAs embedded in the presented PIC. Then, the PXI AI acts only as a pure voltage ADC, without any coupled analog conditioning electronics (as, for example, pre-amplifiers or anti-aliasing filters).

To overcome these two issues, we develop a custom (and purely analog) PCB, which is deeply discussed in the next paragraph.

3.3 Custom Electronics

To compensate for the PXI platform weaknesses, the author of this book has designed a pure analog PCB.

The choices of the Electronic Integrated Circuit (EIC) and their configuration are the result of preliminary pure circuit simulations developed in *LT Spice*. The 3D render of this auxiliary board is shown in Fig. 3.3a, while a photograph of the actually realized device is in Fig. 3.3b.

It has one input connector (68-pin-SCSI, to be coupled to the PXI AO) and one output connector (68-pin-SCSI, to be coupled to the PXI AI) and four bidirectional connectors [60-pin-Flat Flex Cable (FFC), to be coupled to the electrical PC described in the next paragraph].

The DC differential (between ±15 V) power supply is external and provided by means of two BNC connectors. The analog RC-filters and regulators (Texas Instruments $UA7912$, $UA7812$, $UA7905$, $UA7805$ [5, 6]) ensure the correct voltage (and current) at the power supply terminals of the EICs hosted in the PCB, which can be ±12 V or ±5 V.

Some different building blocks, in fact, have been exploited for the design of this board. Right after the input connector [in orange in Fig. 3.3a], there is an array of 8 EICs (Analog Devices, $AD8513$ [7]). To optimize as much as possible the circuit occupation, we choose packages hosting four Operational Amplifiers (for a total of 32). Each Operational Amplifier (OA) (±12 V powered) is in a buffer configuration [negative terminal and output are short-circuited, as in Fig. 3.4a, while each positive input terminal is connected to one of the pin of the input connector]. In so doing, we can not only electrically decouple the PXI-platform

(a)   (b)

Power
Supply
OA stages
INA stages

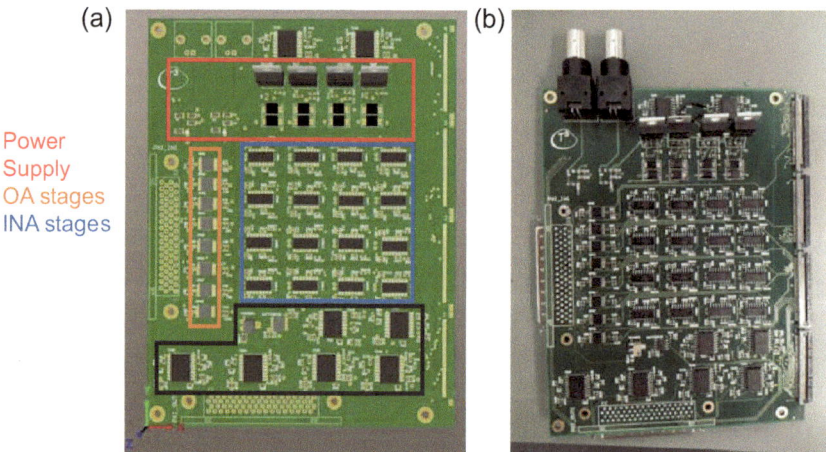

Fig. 3.3 **a** 3D render of the designed PCB. The different stages can be recognized, and highlighted in different colors. **b** Actually realized PCB. The total footprint in 16 cm by 12 cm

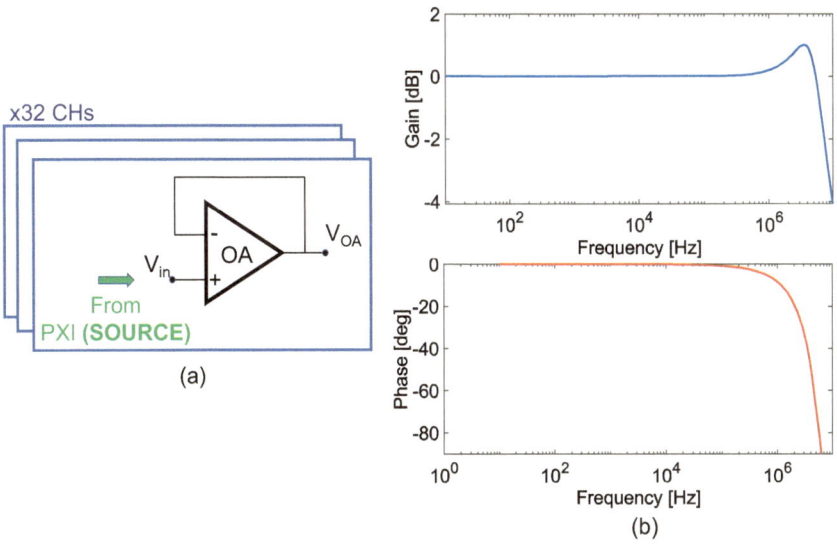

Fig. 3.4 **a** OA in a buffer configuration. The forcing signal is provided by PXI platform and **b** its measured Bode diagram (modulus in blue, phase in orange)

from the probe card (and from the PIC), but we can also enhance the current to be provided to the actuators (approximately 12 mA for heaters and 20 mA for VOAs, in the worst case scenario, according to Table 2.4, and considering 5V voltage drop for heaters and 1.3V for VOAs), while keeping the voltage set by the multichannel PXI AO (whose resolution is limited by the DAC stage, to 300 μV). The maximum current each OA can source is (in safe conditions) 30 mA. The Bode diagram of this first stage is shown in Fig. 3.4b (both modulus and phase).

The control signals applied to the heaters can be updated no more frequently than once every 5 ms (see last paragraph of this Chapter). In addition, we want to keep (electrical) transitions below 5% of this time window (i.e., rise/fall time within 200–250 μs). This means that the electrical bandwidth of the control voltages is below 10 kHz (less than two decades narrower than the bandwidth of the voltage buffer). Again in Fig. 3.4b a peak, at the cutoff frequency can be observed. This is because the open-loop transfer function of the OA has a pole, fairly close (10 MHz) to the frequency at which the closed-loop gain is 0 dB (2 MHz). This reduces the phase margin (<45°).

Each OA output pin is not directly connected to the output connector of the PCB, but we place a 1Ω shunt resistor (R_{Sh}, orders of magnitude smaller than the expected equivalent resistance of the actuators to be driven) between each buffer output and the PC I/O connectors. The voltage drop across each R_{Sh} is read by a suitable INA, configured as in Fig. 3.5a. For the sake of footprint we choose the $INA2126$, from Texas Instruments [8], having two INAs (again ±12 V powered) in the same package. This integrated circuit, having high

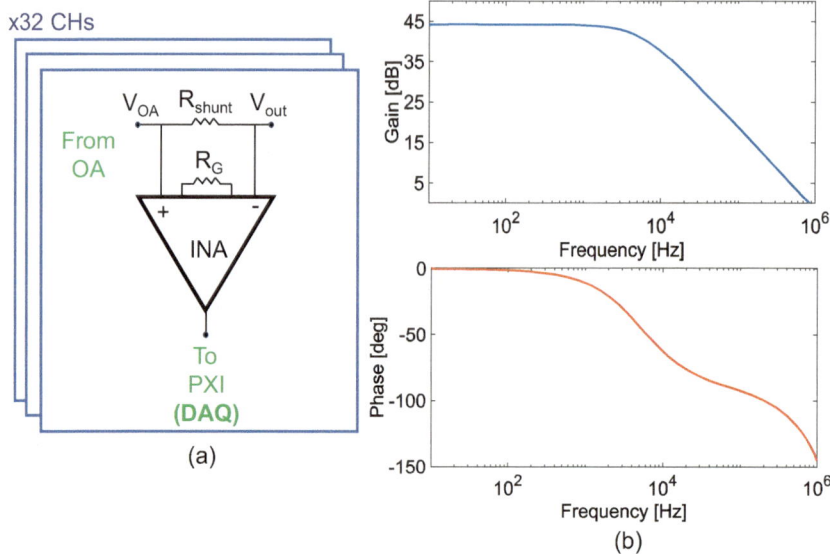

Fig. 3.5 a Circuit configuration of the INA, where exploited R_G is highlighted. **b** Its measured Bode diagram (modulus in blue, phase in orange)

input impedance, does not impair the current flow, but it amplifies the voltage drop across the shunt resistance, which is directly related to the current provided by the buffers. Thus, the output signal of each INA results to be:

$$V_{INA} = G_{INA}V_{Sh} = G_{INA}R_{Sh}I_{Load},\tag{3.1}$$

where V_{Sh} is the voltage drop across the shunt resistor, I_{Load} is the output current of the single buffer and G_{INA} is the gain parameter of this EIC, which deserves further explanation. According to the actual implementation of the internal circuit [8] (composed of two non-inverting amplifying stages):

$$G_{INA} = (1 + \frac{R_1}{R_G})(1 + \frac{R_3}{R_2}),\tag{3.2}$$

where $R_3/R_2 = 4$ and $R_1 = 16\,\mathrm{k\Omega}$[1] (both quantities fixed by circuit designers), while R_G is external with respect to the package and can be chosen by the PCB designer, being the only degree of freedom to set G_{INA}. R_G choice is, indeed, a trade off between the bandwidth of the INA and the gain (in particular the smaller R_G, the higher the gain and the narrower the bandwidth).

---

[1] R_1 is an equivalent resistance, not physical.

For our application we choose $R_G = 500\,\Omega$, leading to $G_{INA} = 165$ which ensures enough gain to detect currents in the range of $\pm 30\,\text{mA}$ (with a minimum current variation of $3\,\mu A$, limited by the intrinsic noise of the circuitry) and, at the same time, enough bandwidth (approximately $9\,\text{kHz}$, sufficient for the dynamics of the involved control signals, as previously discussed and fairly small not to collect excess noise). The Bode diagram (modulus and phase) is shown in Fig. 3.5b.

The INAs' output pins are then connected to the output SCSI connector, which is coupled to the PXI multichannel AI, to be digitized. In so doing, the information about per channel current consumption (or, accordingly, about loads' equivalent resistance) is provided, as a feedback signal, to the PXI controller. This central unit, executing a local control loop, can easily drive the load by setting the electrical power consumption, which is a key point for PIC calibration and testing. In fact, for thermo-optic actuators, the waveguide phase shifts depend only on the temperature shift which is proportional to the heaters' electrical power consumption, not to the applied voltage or to the forced current alone. This means that in the presence of nominally identical waveguides, equipped with non-identical actuators, once the desired phase shift is known (and consequently the heat to be dissipated by the actuator is known), tuning results are much easier.

On the other hand, imposing the same electrical consumption on different actuators belonging to different building blocks, we can state differences among those.

However, the presented PCB is not only devoted to actuators driving but also hosts EIC capable of reading monitor the Ge-PDs embedded in the PIC itself. First, PDs have to be biased. To do so, we exploit one output of the multichannel PXI AO, cascaded with a single OA, in a buffer configuration (just for decoupling reasons). The output of this last OA is then provided (in parallel) to all PDs' cathodes (four in total).

To actually read them, we exploit a logarithmic Trans-Impedance Amplifier (TIA), namely Texas Instruments $LOG2112$ (powered by $\pm 5\,\text{V}$ and again, two of them in the same package [9]). We choose log-TIAs because they offer an extremely high (and easily adjustable, as discussed below) dynamic range and a relatively low noise figure. The actual PCB embeds eight packages (sixteen TIAs), but only two of them (four TIAs) are used. The output voltage of each TIA is a function of the PDs' current (and so of the sensed optical power) as

$$V_{TIA} = [(1 + \frac{R_2}{R_1})nV_T log(\frac{I_{PD}}{I_{REF}})][1 + \frac{R_4}{R_3}]. \tag{3.3}$$

To explain the quantities involved in Eq. (3.3), the circuit scheme of this EIC is necessary and it is reported in Fig. 3.6a. Again, there are two non-inverting amplifying stages (one linear and one logarithmic). The logarithmic stage is implemented by means of an OA in a negative feedback configuration, with two Bipolar Junction Transistors (BJTs) in the feedback branch. Without giving the details, it is clear that the output of the first stage is proportional to the natural logarithm of the ratio between two currents flowing through the collector of the two transistors, which can be converted to the decimal logarithm by scaling the quantity of a factor $n = 2.3$, that is the number appearing in Eq. (3.3). $V_T = 25.85\,\text{mV}$

x16 CHs

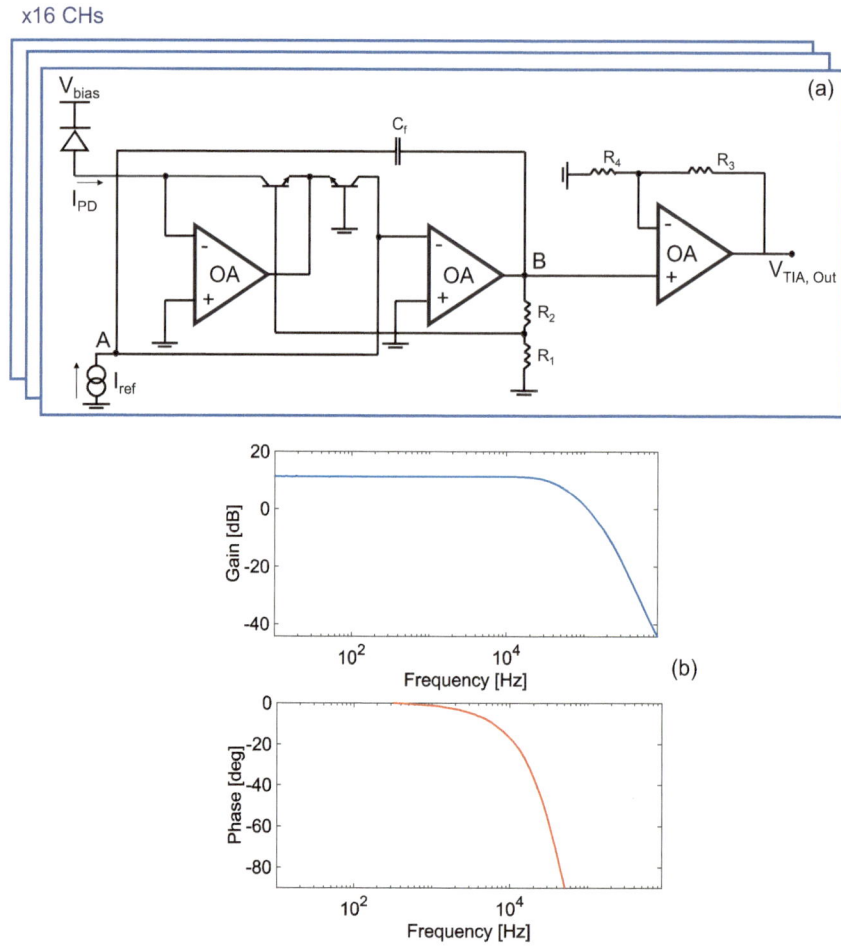

Fig. 3.6 a Circuit configuration of the TIA. **b** Its measured Bode diagram (modulus in blue, phase in orange)

(at room temperature) is the thermal voltage, while ratio $\frac{R_2}{R_1}$ has been chosen by the EIC designer in order to make the product $(1 + \frac{R_2}{R_1})nV_T = 0.5\,\mathrm{V}$. I_{PD} is the current coming out from the single PD's anode (containing information about optical power) and I_{REF} is the reference current, or equivalently, the photocurrent value that makes the TIA's output $V_{TIA} = 0\,\mathrm{V}$, thus setting the middle of the detectable dynamic range. This quantity can be set by the PCB designer by choosing a suitable resistor (R_{REF}) between a reference node (2.5 V fixed by the EIC designer) and virtual ground (the non-inverting input of the first stage OA). We choose $R_{REF} = 249\,\mathrm{k\Omega}$ (the closest value to $250\,\mathrm{k\Omega}$, available on the market), to make $I_{REF} = 10\,\mu\mathrm{A}$ (corresponding to an optical power of $\approx -19\,\mathrm{dBm}$, considering a responsivity of 0.8 A/W [10]). Modifying R_{REF} the median point of the dynamic range correspondingly changes.

Regardless, the logarithmic stage is cascaded with the non-inverting linear amplifying stage, whose gain can be set by the PCB designer through the resistances R_3 and R_4. We choose them to be 1 kΩ and 3 kΩ, respectively, so that $1 + \frac{R_4}{R_3} = 4$. Thus, Eq. (3.3) becomes

$$V_{TIA} = \frac{1}{2}(1 + \frac{R_4}{R_3})log(\frac{I_{PD}}{I_{REF}})[V] = 2log(\frac{I_{PD}}{I_{REF}})[V]. \tag{3.4}$$

With the output voltage swing between -4.5 and 4.5 V and the dark current of the PD (embedded in the PIC under test) $I_{dark} = 150$ nA, range of detectable optical power is between -37 dBm and $+3$ dBm.

Notably, changing the gain of the second stage of the TIA, the detectable dynamic range is modified accordingly. However, the minimum and the maximum optical power values are ultimately given by the PD.

According to Fig. 3.6a, there is one more component to be discussed, the capacitor C_f. Since so-obtained logarithmic amplifiers are usually unstable, there is the necessity to employ a negative feedback capacitor to ensure the stability of the system, in our case between the nodes A and B. Clearly, the larger the capacitor is, the more stable the entire circuit and the narrower the overall bandwidth. For our purposes, we choose C_f equal to 150 pF, ensuring stability (i.e., enough phase margin) and a bandwidth of approximately 46 kHz. This value is compliant with our application since we are dealing with monitor PDs, that have to sense optical power variation in the kHz range. Moreover, by exploiting this configuration, the collected noise is quite low, but the minimum detectable variation of the optical power is limited by PXI-DAC. Its number of bits consents to sample the output voltage of the TIA with a resolution of 0.3 mV (equivalent minimum detectable power oscillation is 0.002 dB). Figure 3.6b shows the transfer function of the entire designed circuit (both modulus and phase).

3.4   Electro-Optical Access of Photonic Integrated Circuits

To carry out flexible and effective testing activities, the I/Os of the DUT must be accessed, to excite the device and recover its response. While for electronic or microwave circuits the access points are only electrical, in Photonics I/O coupling is both optical and electrical. This leverages the complexity of PIC probing, since two different Probe Stations must be exploited, driven by two different mechanical systems, with different features and matching different requirements. Typically, for example, the accuracy of the micropositioners for electrical testing is approximately tens of μm per planar axis [11] (i.e., same order of magnitude of the exposed electrical pads area), while optical coupling (e.g., standard fiber-grating coupler or standard fiber-edge coupler) is indeed more sensitive to the mechanical displacement. For example to have an extra coupling penalty <1 dB (with respect to the optimal position), due to mechanical misalignment, the uncertainty on the fiber tip position has to be within[2]:

- $\pm 3 \,\mu$m, in the case of a grating coupler—standard fiber [12];
- $\pm 1 \,\mu$m, in the case of a edge coupler—small core fiber [12];
- $\pm 1.7 \,\mu$m in the case of a edge coupler—standard fiber (this number has been experimentally measured in our setup).

In the following, an overview of the options to perform effective optical coupling and our proposal to electrically access the PIC under test are reported.

At the moment of writing these lines, an electro-optical probe card is under development (with a proper motion system).

3.4.1   Optical Access

Optical coupling is one of the most time and resource consuming activities in the testing scenario. For some platforms, such as InP, it could be in principle avoided (if there are no other requirements or constraints), exploiting the capability of the platform of the light generation/detection and performing a full-electrical testing.

For other platforms, such as Silicon Photonics, this is not possible and optical coupling must be performed (at least with one input port).

The optical access of a PIC strongly depends on the topology, the spatial position, and the nature of its optical I/Os.

The most common coupling strategies (having different efficiencies, different supported optical bandwidths and different behaviours with respect to the light polarization state) are [12]:

- horizontal coupling (butt-coupling). The fiber-chip coupling occurs on the lateral side of the PIC, with light always propagating in the same plane. In this case the chip and the fiber facets must be of high quality to achieve the best performance. Usually, mode converters are exploited and the coupling efficiency is $>80 - 90\%$ (in the case of lensed fibers [13] [14], while for cleaved fibers the best results in the literature show a coupling efficiency of approximately 50% [12]), the PDL is negligible and the bandwidth is >100 nm (for SOI, approximately 1550 nm). On the other hand, the alignment tolerances are quite strict (sub μm, per axis) [15] and it is difficult to couple a chip before dicing it from its wafer. The typical structures aiding edge-coupling are mode-converters [16], trident architectures [17] or meta-material based couplers [18].

---

[2] These numbers refer to the accuracy per axis, defining the plane parallel to the edge or to the grating coupler, under the hypotheses that the third Euclidean axis (i.e., distance between the fiber facet and coupler itself) is fully optimized.

- vertical coupling. The fiber-chip coupling occurs on the top-surface of the PIC. In all cases, vertical coupling is achieved by means of diffraction gratings, which show a low coupling efficiency (well below 50%), a quite narrow bandwidth (at least one order of magnitude below the typical edge coupling structures) and an evident polarization sensitivity [19]. However, the coupling tolerances are less tight (a few μm per axis), and this approach is, by definition, suitable for WLT.

Regardless of the coupling strategy, a single cleaved fiber(s) (moved by sub-micron micropositioners) is used to optically access the PIC, if the number of I/Os is limited (i.e., 1–2). In contrast, for a higher number of optical ports, fiber blocks (in case of horizontal coupling) or transposers (in the case of vertical coupling) are successfully exploited for complex multiport PIC testing. A common value for the clearance between two optical accesses is 127 μm.

However, in principle, only the use of transposers effectively allows for WLT, while to optically access a PIC through a fiber block, preliminary dicing is needed.

Nevertheless, in the recent years suitable structures to perform edge coupling at the wafer level have been proposed [20], exploiting the trenches etched among the different circuits hosted in the same substrate [21]. Among these we can find those presented in [22], perfectly compliant,[3] from the geometrical point of view, with the trenches in silicon photonic wafers. Moreover, these structures show a very good coupling efficiency (theoretically >90%), if put in the correct position, with a tolerance of ≈1 μm per axis.

To achieve these performances, from a mechanical standpoint, extremely precise piezo-actuators are often used along with active alignment routines. Recently other strategies have also been demonstrated, for example based on Micro-Electro-Mechanical Systems (MEMS) [23].

However, these approaches are quite time and resource consuming (duration can be in the order of seconds, with 50 nm-accuracy positioners), and definitely constitute a limit for the testing throughput.

To overcome these issues some solutions based on passive fiber-chip alignment have been provided, such as those in [24, 25].

3.4.2   Electrical Probe Card Exploitation and Characterization

Usually, in many works present in the literature, wire bonding is the electrical interface between a PIC and the external world, particularly in the presence of a high density of electrical pads. Wire bonding a chip onto a motherboard, indeed, ensures stability, reliability and, in general, a high-frequency cut-off (hundreds of MHz, at least). However, this approach, by definition, is not suitable for (high volume) testing, especially at the wafer level.

The same limitations arise for the flip-chip coupling.

---

[3] Width of trenches is usually >50 μm.

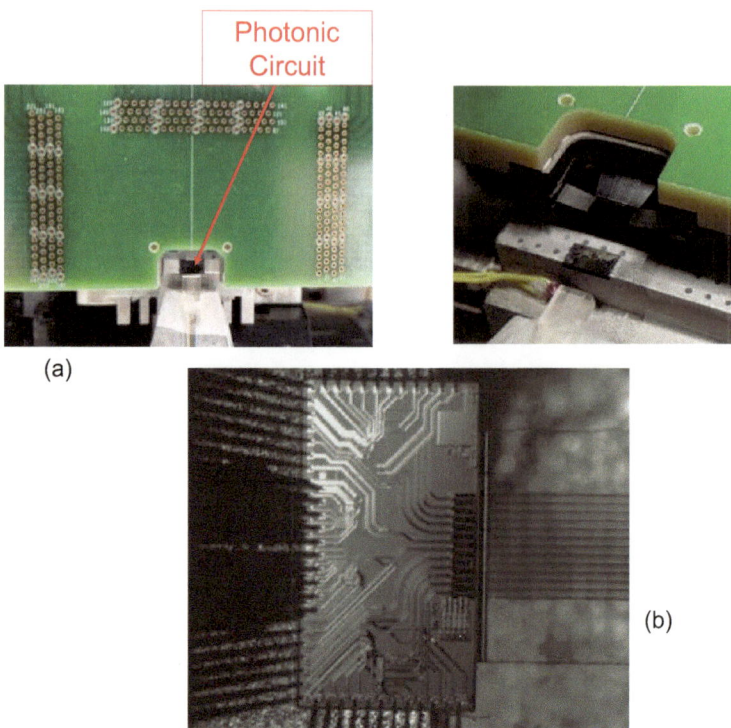

Fig. 3.7 a Photographs of the PC, placed above the PIC (the four-channels TOADM). **b** Microphotograph of the PIC, surrounded by the needles and optically contacted by an external fiber block

Conductive needles, instead, are more flexible and allow an electrical access of the DUT,
without destroying its pads. However, they are usually limited in number (2–4 per setup).

Since in the fields of electronics and microwave (but also recently in photonics), Probe
Cards are successfully and advantageously exploited, for testing purposes [26], we design
a suitable one, shown in Fig. 3.7a, to test the PIC previously described.

With respect to the PCs for photonic testing applications already reported in the literature
[27], that proposed in this book can handle a higher number of electrical channels (52 pads
need to be contacted for the described PIC, but we foresee up to 202 electrical I/Os) and it
is ready to be integrated with suitable 3-D printed structures for optical edge coupling [22].

PCs are essentially PCBs with needles (whose length is $240 \pm 15\,\mu$m and whose diameter
is $23 \pm 3\,\mu$m) collected in a spider-like-structure and properly glued underneath. The needles
are made of Tungsten Rhenium Alloy, a suitable material to minimize the parasitic resistance
with the aluminum pads, with which the PIC under test is provided. Their relative positions
are matched with the pad layout of the DUT and, when in contact they conduct the electrical
signal into the proper point of the circuit. The hosting PCB we design is purely passive,

conveying all the signals coming from the electronic board discussed in Sect. 3.3, from the standard connectors (in our case four 60-pin-FFC) to the needles themselves. As can be observed in Fig. 3.7b, we put needles along the three sides (coherently with the PIC electrical layout), leaving one side completely open, for optical coupling, which occurs, at the moment, by means of a standard fiber block (composed of twelve SMF-28 fibers,[4] 127 μm spaced, matched with suspended couplers of the DUT).

Thus, the described PC allows for flexible electrical access of the PIC, simultaneously contact all its electrical pads. This enables parallel testing of all the electrical sub-devices embedded into the PIC itself and for the execution of the calibration routines, preliminary to the optical testing. To achieve effective electrical conduction, the PC must be carefully positioned, controlling all the axes (x, y, z, roll, pitch and yaw) of the mechanical positioner. The encoders mounted on the axis and pattern recognition [Optical Character Recognition (OCR)[5]] allows for precise positioning of the PC on the horizontal plane (identified by the x-y axis), well within each electrical pad area (90 μm by 90 μm), but to perform a proper touchdown (i.e., needle-pad mechanical contact) closed-loop control of the pushing force is needed. The cantilevers of the PC, in fact, require a contact force per unit length compression of 0.1 g/μm/pad (and typical compression values are ≈100 μm). This is a key point. In the volume testing scenario, the number of touchdowns per unit time may be very high, and the needles of the PC may suffer from aging and weathering. Thus relying only on the relative vertical distance PIC–PC could be misleading, inducing a non-perfect needle-pad contact. By means of a piezo-resistive sensor, which guarantees force estimation with sufficient accuracy, combined with vertical distance estimation, proper touchdown is always performed.

The sensitivity of the performances on z-alignment was evaluated through a suitable (resistive) test structure, by repeating the touchdown 10 different times. Each time resistance is read by means of a Digital Multimeter (DMM)-Keithley 2450 characterized by a 16-bit digitizer, with a programmable measurement range spanning from 10 mΩ to 10 MΩ with DC-measurements. The outcomes are shown in Fig. 3.8a. In just one case the pushing force is not sufficient and the resistance is not correctly read.

Once putting the PC in its own nominal contact condition, we characterize it, from the electrical point of view, focusing on the needles-pads parasitic resistances and on the parasitic capacitance between two consecutive needles. The needle-pad contact resistance (for all the cantilevers) has been measured exploiting a suitable test structure and again the DMM. The outcome of the experiment is reported in Fig. 3.8b.

The mean value is 1 Ω, with a 3σ = 0.5 Ω and it can be considered a negligible contri-bution, if cascaded with the PIC actuators, we are dealing with. However, as this parasitic

---

[4] Fiber block with UHNA-7 fibers was not available at the time of these measurements, so we had to use this one. The non-negligible drawback is, of course, the coupling efficiency. SMF fibers give an extra penalty of approximately 4 dB per port.

[5] Artificial Vision is also used to perform a preliminary Wafer-Level Inspection, to reduce, in case of detected issues, the number of potential DUTs [28].

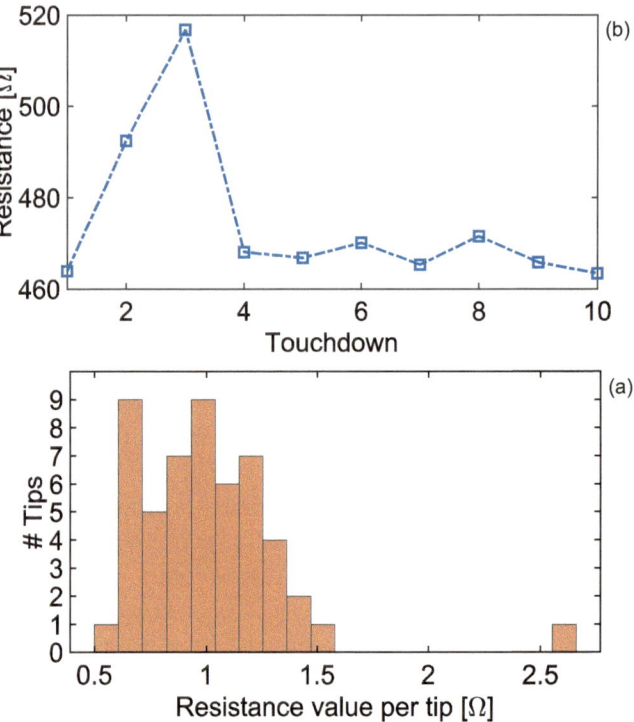

Fig. 3.8 a Measured test resistance for ten different touchdowns. **b** Distribution of the needle-pad parasitic resistances values

resistance is also present between the actuators common node and ground, it may affect the frequency behavior of the PC, as demonstrated below.

According to the electrical layout of the PIC, the needles are approximately $100\,\mu m$ apart, one from another. Thus, the parasitic capacitances between two consecutive electrical channels, may limit the maximum operative frequency of the PC.

To quantitatively analyse this impact, we exploit a Vectorial Network Analyzer (VNA) (E5061B, from Keysight). The VNA is set up to sweep the forcing signal (whose amplitude is 30 mV) between 10 Hz and 1 MHz, with 50 points per decade.

Three needles (not touching any circuit) are considered. Two of them (labeled as "A" and "B") are $100\,\mu m$ apart (that is the minimum distance between two cantilevers of the probe card). The third one (labeled as "G") is much farer (i.e. $>1\,mm$). The goal is the evaluation of the mutual coupling between needles "A" and "B". The forcing signal ($V_f(f)$) is applied between "A" and "G", while the voltage drop induced between "B" and "G" ($V_c(f)$) is monitored through the test-port of the VNA. The amplitude of this mutual coupling ($|H(f)| = |V_c(f)|/|V_f(f)|$) is shown in Fig. 3.9c (blue curve). The presence of a zero at

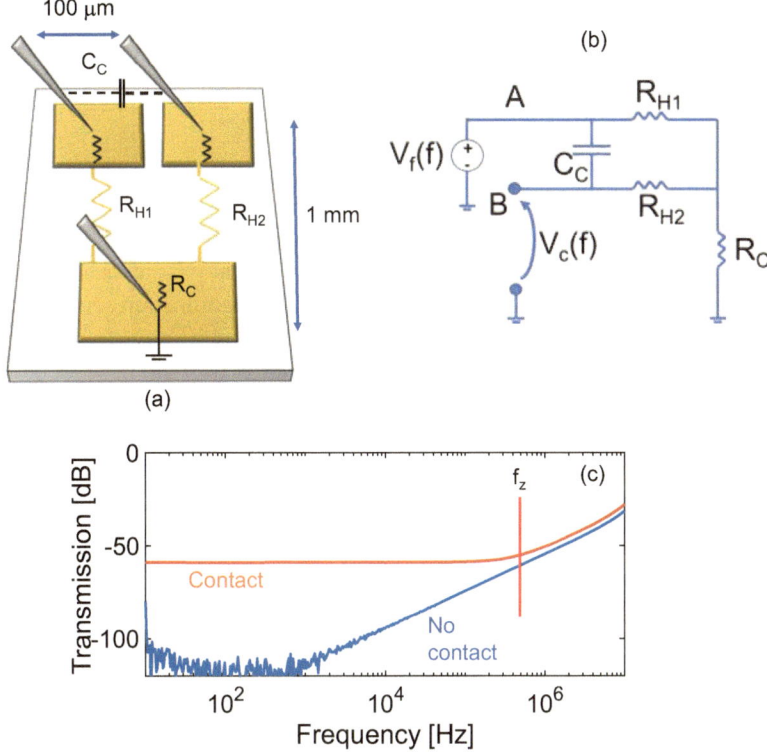

Fig. 3.9 a Sketch (not in scale) of the considered situation, with three needles in contact (needles are 90 μm by 90 μm, not in scale) and **b** its circuit representation. **c** Measured electrical crosstalk between closer needles

the origin of the plot is indeed clear so the mutual electrical coupling between "A" and "B" is.

Now, let us consider the same set of needles, but in contact with the photonic circuit under test. More specifically, between the needles "A" and "G" there is a resistor $R_{H1} = 400\Omega$, while between "B" and "G", there is a resistor $R_{H2} = 450\Omega$. "G" is the common ground of the PIC. The situation is sketched in Fig. 3.9a, while the equivalent circuital scheme is in Fig. 3.9b. Strictly speaking the goal is the evaluation of the mutual capacitance (C_C) between "A" and "B". The contribution of the parasitic contact resistor is shown (as $R_C = 1\Omega$). Again, $V_f(f)$ is applied between "A" and "G", while $V_c(f)$ is collected between "B" and "G".

The computation $|H(f)| = |V_c(f)|/|V_f(f)|$ is performed and the result is shown in Fig. 3.9c (orange curve). For low frequencies the isolation between nodes "A" and "B" is approximately 60 dB, while at higher frequencies, it decreases. In fact, according to Fig. 3.9c (orange curve), the measured transmission increases for frequencies higher than 360 kHz

(f_z, where there is a zero). Exploiting this quantity, an estimation of C_c can be given. The theoretical computations lead to

$$f_z = \frac{1}{2\pi C_C (\frac{R_{H1} R_{H2} + R_{H1} R_C + R_{H2} R_C}{R_C})}, \tag{3.5}$$

and thus C_C has to be approximately $\approx 3\,\mathrm{pF}$, in line with expectations. The working frequency of hundreds of kHz, being larger than the bandwidth of the EICs of the custom analog board, could be considered sufficient for the electrical measurements described in the following sections and for the execution of the control algorithm presented in Chap. 4.

3.5 Electrical Measurements

The simultaneous contact of all the PIC exposed pads, allows for properly driving and testing of all the electrical sub-devices embedded in the photonic structure. Nominally, with the specifications given above in terms of the employed material (Tungsten Rhenium Alloy) and the geometrical dimensions (23 μm diameter), each needle can handle more than 30 mA of current, a fair amount for silicon photonic circuit actuators.

In this context with the word "testing" we intend to collect of all the current-voltage (I–V) characteristic curves of the embedded phase shifters and attenuators and estimate the sensors' noise. The PDs are reverse biased, with a DC-voltage of 2.4 V, which is the optimum working point for the Ge-detectors, in this platform [10]. To acquire parallel I–V curves, by using the AO PXI board, we generate 32 (i.e., number of actuators to be tested) discrete voltage steps, from 0 V to 5 V (with $\Delta V = 0.1$ V). Each step has a duration of 1 ms (the bandwidth of the INA is $<10\,\mathrm{kHz}$, and the transients are $>100\,\mu\mathrm{s}$), so the ramp has an integral duration of 50 ms.

Thanks to the chain of EICs described in Sect. 3.3, a sufficient amount of current can be delivered to the load, and this quantity can be easily read and digitized, using the analog PCB and AI PXI board. By synchronizing AO and AI modules, the forced voltage and the induced current can be easily correlated in time. Averaging the digitized current and voltage samples, over the duration of the single step, the single point of the I–V curve is obtained, and we can build the entire characteristic line. To test all the sub-devices of the considered PIC in fair conditions we perform the test in four consecutive steps:

- I–V curve acquisition of heaters above the top MZI, Ring 2 and Ring 4, in parallel for the four filters,
- I–V curve acquisition of the heaters above Ring 1, Ring 3 and the bottom MZI, in parallel for the four filters,
- I–V curve acquisition of the p-n junction surrounding Ring 2 in parallel for the four filters,

- I–V curve acquisition of the p-n junction surrounding Ring 3 in parallel for the four filters.

This hybrid parallel-serial approach is four times longer than the full parallel approach, but in so doing we eliminate the thermal coupling, which may change the behaviour of the diodes and the resistors embedded in the PIC, making the measurements unreliable (the thermal coupling, instead, can be considered negligible among non-consecutive devices, as will be extensively discussed in the next section).

The set of measured I–V curves (for the heaters and p-i-n junctions) is reported in Fig. 3.10a. There are heaters controlling the same building block, from the PIC topology point of view, which are inherently different. This demonstrates for the importance of a heater control based on the consumed electrical power, or equivalently, to have simultaneous access to the information about the current absorbed by the load.

To quantitatively evaluate these differences, data are processed and in particular, we compute the heaters' resistances and the threshold voltages of the p-i-n junctions. The first quantity is computed as follows:

$$R_j = \frac{1}{N-1} \sum_{i=2}^{N} \frac{V_{j,i} - V_{j,i-1}}{I_{j,i} - I_{j,i-1}},$$ (3.6)

where j is an index to enumerate the heaters, and V_i and I_i are the I–V curve points, composed of $N = 50$ points in total.

The second is computed by using the tangent method, as shown in Fig. 3.10b.

Furthermore, during the I-V curve measurements, the TIAs' output voltages are sampled and averaged, obtaining a measure of the PDs dark current and the amplifiers' noise figure [Fig. 3.10c].

Once, the heaters and the p-i-n junctions are ensured to work properly, a single monochromatic light source is coupled (through a 6 dB splitter) to the four Add ports of the DUT, and integrated VOAs are turned on.

The light beam is progressively attenuated by means of an external voltage-controlled VOA (from 0 dB to 30 dB extra attenuation, with steps of 3 dB, every 10 ms, with the VOA electrical bandwidth limited to 1 kHz). Simultaneously, the TIAs' output voltages are acquired. In so doing, the optical power-TIA voltage for each PD can be obtained [the curve for one single PD-TIA pair is Fig. 3.10d].

In the case of substantial and unexpected insertion loss (due to a PD malfunction, a waveguide interruption or a broken edge-coupler), the test is considered failed.

Thus, from this electrical analysis, we can not only state whether an electrical sub-device is working or not (i.e., if it is compliant with specifications or not), but we can infer the statistical distributions of the computed relevant parameters, as shown in Fig. 3.11a–c, for the heaters' equivalent resistances, the VOAs' threshold voltages, and the PDs' (with TIAs) responsivity. The total duration of electrical testing is 300 ms (200 ms for I–V curves and 100 ms for PD responsivity). A strategy to enhance throughput is reported in Appendix A.

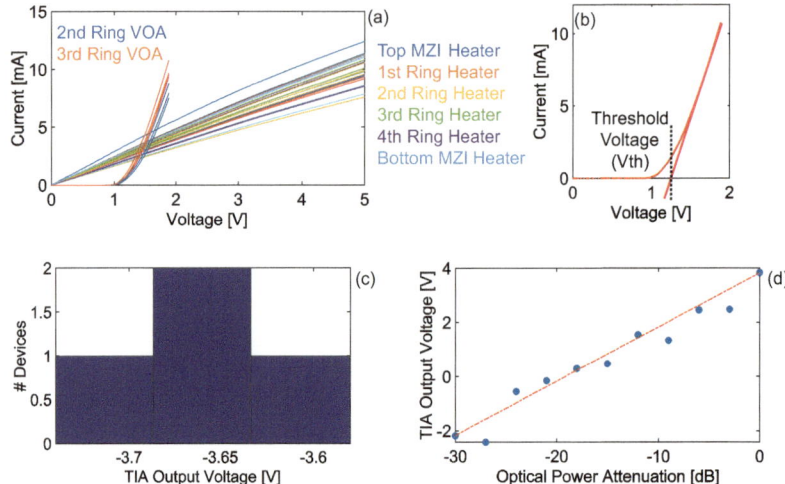

Fig. 3.10 a I–V curves for all the actuators embedded in the four channel TOADM. The colors are matched to the same building block in the four structures. **b** I–V curve of a p-i-n junction. Its tangent (in the maximum linearity region) is exploited to compute the device threshold voltage. **c** Distribution of the four TIA output voltages, in dark conditions. **d** Optical power-TIA Output Voltage characteristic for one PD

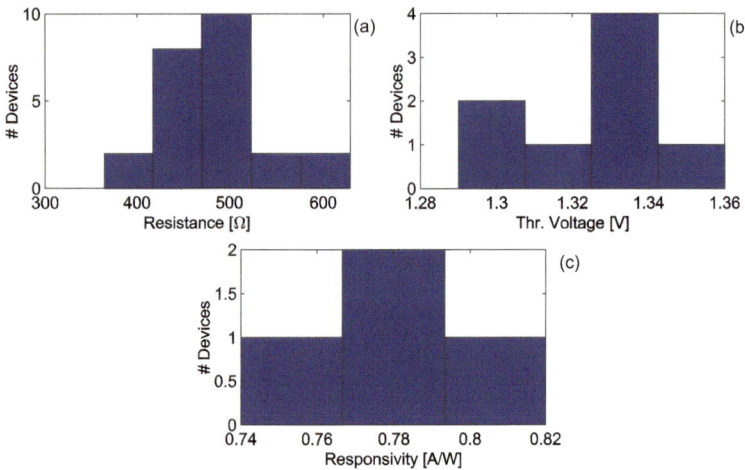

Fig. 3.11 Distribution of heaters' resistances (**a**), p-i-n junctions threshold voltages **b** and PDs responsivity

To conclude, we believe that there is no photonic testing without effective electrical testing (if the sensors or the actuators do not properly work, the quality of the photonic chip is severely impaired) and it constitutes an important precondition for optical testing, giving some useful information for PIC calibration, as stated in the next paragraph.

3.6   Thermal Crosstalk Evaluation

As mentioned in Chap. 2 and in the previous paragraphs of this Chapter, the DUT is equipped with actuators, relying on the thermo-optic effect. This kind of actuator is well established in Photonics and allows for a very precise phase tuning [29] but they present a couple of drawbacks: a quite slow time response (in the microsecond time range) and the thermal crosstalk. In particular, the effect of the latter may be detrimental, slowing down or even impeding the execution of a control process. An effective cancellation of the thermal crosstalk has been already proven in [30, 31], by using the TED, which works as follows (Fig. 3.12).

Let us consider a generic circuit made of four elements, sketched in Fig. 3.14a. Due to the thermal crosstalk, the phase coupling, when the j-th heater is activated, induces a phase shift of $\delta\phi_j$ on the element that directly controls and a spurious phase shift $\delta\phi_i = \delta\phi_j T_{j,i}$ on the i-th element, where $T_{j,i}$ is the phase coupling coefficient. All the generic $T_{j,i}$ can be collected in a matrix $\mathbf{T}$. Hence, $\mathbf{T}$ describes the relation with a desired set (vector) of phase shifts ($\boldsymbol{\delta\phi}$) and the actual one ($\boldsymbol{\delta\tilde{\phi}}$), as

$$\boldsymbol{\delta\tilde{\phi}} = \mathbf{T}\boldsymbol{\delta\phi}. \tag{3.7}$$

Equation (3.7) can be conveniently rewritten as

$$\boldsymbol{\delta\tilde{\phi}} = \mathbf{P}\mathbf{T_D}\mathbf{P}^{-1}\boldsymbol{\delta\phi}, \tag{3.8}$$

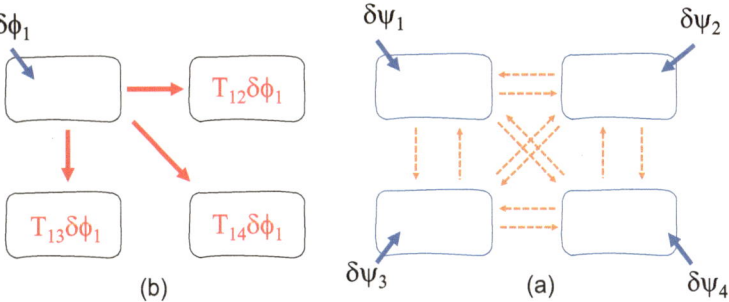

Fig. 3.12 a Thermally coupled system scheme. When a certain phase shift is imposed on an element, those in the vicinity experience spurious phase shifts. **b** All the phase shifts are properly weighted and applied simultaneously

where $\mathbf{P}$ is composed of the column eigenvectors of $\mathbf{T}$ and $\mathbf{T_D}$ is a diagonal matrix containing the corresponding eigenvalues. From Eq. (3.8), the transformed coordinates $\delta\Psi = \mathbf{P}^{-1}\delta\phi$ can be defined and the obtained system is thermal-coupling free, because $\delta\dot{\Psi} = \mathbf{T_D}\delta\Psi$. Consequently, all the heaters must be simultaneously controlled [as shown in Fig. 3.14b], with appropriate weights, coming from the eigensolutions of the thermally coupled system. Weighting the set-points of the heaters with these coefficients, any control approach on any circuit topology can be executed, including that described in the next chapter, based on feedback iteration.

Hence, the knowledge of matrix $\mathbf{T}$ becomes crucial and noteworthy TED can be advantageously implemented even when $\mathbf{T}$ is partially known or approximated.

Typically, for reasonably complex devices, this matrix can be obtained by means of suitable optical measurements or heuristic computation (i.e., accounting only the topology for the PIC). However, the former approach cannot always be performed, while the latter might be unreliable. In this work, we have chosen pure electrical measurements, exploiting the heaters embedded in the PIC as temperature sensors. This approach represents a trade-off between accuracy and actual practicability. Using the electrical testing system widely described in the previous paragraphs, when a heater (let us call it an "aggressor") is turned on and dissipates a certain amount of power (e.g., the nominal power to achieve a 2π phase shift, discussed in Chap. 2), the I–V curves of the neighboring actuators (let us call them "victims") are measured, one at a time. By stating the change of victims' electrical resistances, we are able to estimate the temperature shift of the waveguides under them (assumed equal to the temperature shift sensed by the resistor, $\Delta T_{vict,i}$). Thus, the single entry of $\mathbf{T}$, $T_{i,j}$, by definition is:

$$T_{i,j} = \frac{\delta\tilde{\phi}_i}{\delta\phi_j} = \frac{\Delta T_{vict,i}}{\Delta T_{agg,j}}, \tag{3.9}$$

where $\Delta T_{vict,i}$ is the equivalent spurious temperature shift induced on the i-th victim and $\Delta T_{agg,j}$ is the temperature shift imposed by the j-th aggressor on the waveguide underneath, recalling that the generic phase shift is

$$\delta\phi = \frac{2\pi(\delta n_{eff})L}{\lambda_0} \tag{3.10}$$

where $\lambda_0 = 1550\,\text{nm}$ and L is the length along which the variation in the effective refractive index (δn_{eff}) occurs. That variation is

$$\Delta n_{eff} = K_{eff}\Delta T, \tag{3.11}$$

where

$$K_{eff} = \sum K_i \Gamma_i \approx K_{th,Si}\Gamma_{Si}. \tag{3.12}$$

K_{eff} is the effective thermo-optic index of the platform (namely the linear combination of the various materials [32]) and Γ_i is the confinement factor of the optical mode in the single material. $K_{th,Si}$ is the thermo-optic coefficient of silicon (waveguide core), which is

Table 3.1 $\Delta T_{2\pi}$ for the different building blocks

Building block	FSR [nm]	$\Delta T_{2\pi}[K]$
Top MZI	6.7	84
Ring 1	6.7	84
Ring 2	11.2	140
Ring 3	9.6	120
Ring 4	8.0	100
Bottom MZI	8.0	100

$1.86 \cdot 10^{-4}\,\mathrm{K}^{-1}$ (approximately 300 K), ΔT is the generic temperature shift and $\Gamma_{Si} = 0.75$ is the confinement factor computed for the considered waveguide cores (at the wavelength of interest). If the building block controlled by this kind of actuator shows a frequency selective spectral response, it is shifted (in the wavelength domain) by the following amount

$$\Delta \lambda = \frac{\Delta n_{eff} \lambda_0}{n_g} = \frac{K_{eff} \Delta T \lambda_0}{n_g}, \tag{3.13}$$

considering $n_g = 3.89$, $\Delta\lambda/\Delta T$ is around 75 pm/K.

In the previous equations, the substrate expansion contribution and the SiO_2 (cladding) thermo-optic effect are neglected, being 100 and 10 times lower than the Si thermo-optic effect, respectively [33]. Moreover, the optical mode is well confined in the Si-nanowire core.

Here, $\Delta T_{agg,j}$ is considered to be the temperature to impose a 2π phase shift on the transfer function (one FSR in the wavelength domain) of the j-th sub-device tuned by the heater aggressor ($\Delta T_{agg,2\pi,j}$), which is

$$\Delta T_{agg,2\pi,j} = \frac{n_g FSR_j}{K_{eff} \lambda_0}. \tag{3.14}$$

Thus knowing the FSR of the building blocks of the PIC under test, each $\Delta T_{agg,2\pi,j}$ can be easily computed and the actual numbers are reported in Table 3.1. $\Delta T_{agg,2\pi,j}$ is imposed by making the aggressor heater dissipating the power to achieve a 2π shift.

In this work, to sense the generic ΔT we exploit Titanium Nitride (TiN) heaters, whose resistance-temperature dependence has been experimentally measured as follows. While intentionally changing the temperature of a test photonic chip hosting a single heater (whose dimensions are almost the same as those of the DUT) from 10 °C (283 K) to 40 °C (313 K) (using a TEC), we measure the resistance change (passing from 399.7 Ω to 405.1 Ω), as reported in Fig. 3.13a. In this range the resistance-temperature dependence appears to be

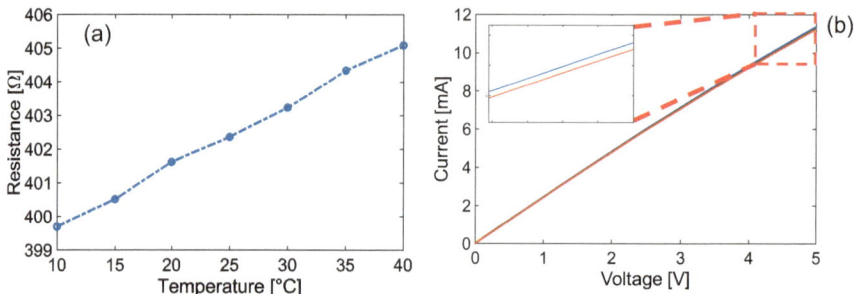

Fig. 3.13 **a** Resistance-temperature dependence, measured on a test structure. **b** I–V curve of the victim not affected (blue curve) and affected (red curve) by the thermal crosstalk

linear, with $\Delta R/\Delta T = 180\,\mathrm{m\Omega/K}$, which is well detectable with the previously described electronic system.[6]

As an example, we considered a single TOADM, kept at 25°C by means of a TEC. In Fig. 3.13b two I–V curves referring to the heater above the first ring (of one of the four filters) are reported, when the heater above the second ring is turned off (blue curve) and when it dissipates the power to induce a 2π phase shift on MRR #2 (orange curve)—i.e., $\Delta T_{agg,2\pi} = 140\,\mathrm{K}$. The resistance variation is indeed evident, being approximately 4.8 Ω, which corresponds to a temperature shift of 27 K.

Thus, knowing the electrical characteristics of the heaters and repeating the procedure for each pair of actuators, the induced temperature shifts on the victims by the aggressors are assessed. The induced temperature matrix and the thermal crosstalk matrix for one channel of the accounted DUT are measured and calculated, and shown in Tables 3.2 and 3.3, respectively.[7]

All the main diagonal elements of **T** are, by definition, equal to 1, while the off-diagonal elements are quite high, but still reasonable. The crosstalk between the closest elements is, in fact, between 8% (Tuneable Coupler—Ring) and 20% (Ring-Ring). Instead, the crosstalk among non consecutive elements and among different filters appears to be negligible (at least with our read-out electronics).

To prove the effectiveness of the proposed approach to compute **T**, we perform optical measurements on suitable test structures, such as that in Fig. 3.14a. This structure (whose temperature is always 25°C) is composed of two heaters whose dimension, position and shape are exactly the same as those of the first two of the actual filter and a single MRR (whose radius is equal to the radius of the first ring of the filter, 14.37 μm) is placed below

---

[6] Minimum detectable temperature shift with this setup is 0.5 K. Even if this quantity appears to be coarse, it enables a rough estimation of the matrix **T**, enough for the TED implementation.

[7] The matrix is not symmetrical since the building blocks are inherently different.

Table 3.2 Induced temperature shift (in K degrees) on the i-th victim, when the j-th aggressor experiences $\Delta T_{2\pi}$

	Top MZI	Ring 1	Ring 2	Ring 3	Ring 4	Bottom MZI
Top MZI	84	8.1	0	0	0	0
Ring 1	8.2	84	16.1	0	0	0
Ring 2	0	19	140	19.2	0	0
Ring 3	0	0	23.2	120	23.0	0
Ring 4	0	0	0	19.8	100	9.8
Bottom MZI	0	0	0	0	9.9	100

Table 3.3 Thermal crosstalk matrix (T)

	Top MZI	Ring 1	Ring 2	Ring 3	Ring 4	Bottom MZI
Top MZI	1	0.0964	0	0	0	0
Ring 1	0.0976	1	0.115	0	0	0
Ring 2	0	0.226	1	0.160	0	0
Ring 3	0	0	0.166	1	0.231	0
Ring 4	0	0	0	0.165	1	0.098
Bottom MZI	0	0	0	0	0.099	1

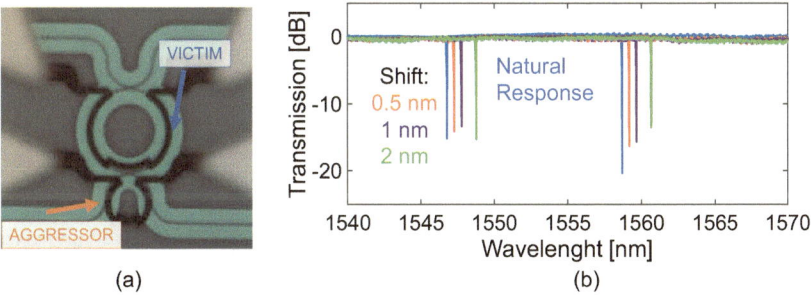

(a)   (b)

Fig. 3.14 a Test structure exploited for the optical evaluation of the thermal crosstalk, constituted by two heaters and one MRR, whose resonance drifts **b** in the wavelength domain due to the temperature shift induced by the aggressor

the first heater. We turn on the second one (aggressor) and we measure, by means of a TLS synchronized with an OSA the wavelength shift of the MRR resonance, as shown in Fig. 3.14b.

When the aggressor dissipates 12, 24 and 48 mW (which corresponds to induced phase shifts of 1.25, 2.5 and 5.2 rad onto the waveguide underneath if present), the victim's resonance is displaced by 0.5, 1 and 2 nm, respectively. This corresponds to a spurious phase shift of 0.26, 0.52 and 1.04 rad and thus crosstalk of approximately $20 \pm 2\%$, recalling Eq. (3.9).

The results obtained with the optical measurements are well in line with those of electrical measurements, ensuring the reliability of our approach. Notably, with this method, we avoid the use of bulky and (usually) slow instruments, such as the pair TLS-OSA, obtaining **T** for the entire PIC in a few hundred milliseconds. In fact, 10 different pairs of heaters (per filter) must be considered, and each I–V curve measurement takes approximately 50 ms (since there is no crosstalk among the different channels, the measurements for the four filters can be performed in parallel), for a total amount of 500 ms. Furthermore, the approach also works if the thermo-optic actuators are controlling the non-frequency selective building blocks (such as waveguides or balanced MZIs), for which optical characterization of thermal crosstalk would be ineffective.

3.6.1   Thermal Crosstalk Time Evolution

The thermal coupling coefficients evaluated so far are computed at the steady state. However, the thermal field propagation from the aggressor to the victim is not instantaneous and to calculate **T** one has to wait for the transition exhaustion and also to effectively apply TED method. This limits the number of corrections per unit time that can be applied to the heaters. Thus, it is important to know the time constants involved in this scenario, to state the minimum time frame needed per single tuning of the DUT.[8]

To do so, we consider a structure such as that shown in Fig. 3.14a. We drive the aggressor with a square wave, whose repetition rate is 100 Hz (with 50% duty cycle), between 0 V and 5 V. The rise/fall time is $<1\mu$s. At the same time, the resistance variation of the victim (biased with a DC voltage equal to 1.5 V) is measured and sampled at a sufficiently high rate (10 MHz).

The outcome of this experiment is shown in Fig. 3.15. The waveform is well approximated by a single pole step response, with $\tau = 1.1$ ms (transition is exhausted in $5\tau = 5.5$ ms).

This number is well in accordance with the following [34] (single direction thermal field propagation):

$$\tau_v = \tau_a (\frac{L_v}{L_a})^2, \tag{3.15}$$

where L_a is the distance between the aggressor and its mainly heated waveguide, and L_v between the aggressor and victim waveguide. τ_a is the time constant of the thermo-optic

---

[8] Note that this is not a testing stage, instead it is more related to the characterization of the chip. The measurement was carried out by a proper setup, composed of a fast arbitrary waveform generator and sensitive Source and Measurement Unit(SMU).

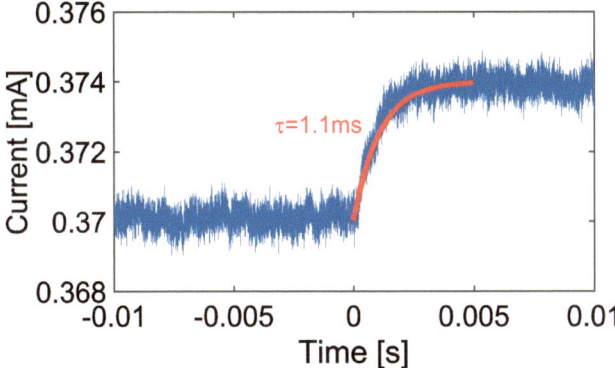

Fig. 3.15 Current variation of the victim, when the aggressor is driven by a square wave, with rise/fall time $<1\,\mu$s. The waveform is well approximated by a single pole step response

effect induced by the aggressor on its main waveguide.[9] τ_v is the time constant of the thermal crosstalk between aggressor and victim.

In the case of the proposed photonic chip, $L_a = 700$ nm, $\tau_a < 10\,\mu$s and $L_v < 10\,\mu$m.

According to Eq. (3.15) this time constant depends on the average distance between the aggressor and the victim, but exploiting the described system, the crosstalk among non-adjacent rings is negligible.

References

1. PXIe-8880, *PXI Express*. NI, 2020.
2. PXIe-1092, *Features*. NI, 2020.
3. PXIe-6738, *32-Channel High-Density Analog Output*. NI, 2020.
4. PXIe-6355, *X Series Data Acquisition*. NI, 2020.
5. UA7800, *Positive Voltage Regulators*. Texas Instruments, 2006.
6. UA7900, *Negative Voltage Regulators*. Texas Instruments, 2006.
7. AD8513, *Precision, Very Low Noise, Low Input Bias Current, Wide Bandwidth JFET Operational Amplifiers*. Analog Devices, 2018.
8. INA2126, *INAx126 MicroPower Instrumentation Amplifiers*. Texas Instruments, 2022.
9. LOG2112, *Precision Logarithmic and Log Ratio Amplifier*. Texas Instruments, 2005.
10. "PDK of AMF, Advanced Micro Foundry, Singapore, [online]. available: www.advmf.com."
11. V. A. M. Bushnell, *Essentials of Electronic Testing for Digital, Memory and Mixed-Signal VLSI Circuits*. Springer US, 2006.
12. R. Marchetti, C. Lacava, L. Carroll, K. Gradkowski, and P. Minzioni, "Coupling strategies for silicon photonics integrated chips (Invited)," *Photonic Research*, vol. 7, no. 2, 2019.

---

[9] We can assume that the distance between the aggressor and victim waveguide and the distance between the aggressor and the heater above the victim waveguide (i.e., thermal probe) are approximately the same.

13. C.-H. Lin, S.-C. Lei, W.-H. Hsieh, Y.-C. Tsai, C.-N. Liu, and W.-H. Cheng, "Micro-hyperboloid lensed fibers for efficient coupling from laser chips," *Optics Express*, vol. 25, no. 20, 2017.

14. Q. Fang, T.-Y. Liow, J. F. Song, C. W. Tan, M. B. Yu, G. Q. Lo, and D.-L. Kwong, "Suspended optical fiber-to-waveguide mode size converter for Silicon photonics," *Optics Express*, vol. 18, no. 8, 2010.

15. X. Mu, S. Wu, L. Cheng, and H. Fu, "Edge Couplers in Silicon Photonic Integrated Circuits: A Review," *Applied Sciences*, vol. 10, no. 4, 2020.

16. K. Kasaya, O. Mitomi, M. Naganuma, Y. Kondo, and Y. Noguchi, "A simple laterally tapered waveguide for low-loss coupling to single-mode fibers," *IEEE Photonics Technology Letters*, vol. 5, no. 3, 1993.

17. N. Hatori, T. Shimizu, M. Okano, M. Ishizaka, T. Yamamoto, Y. Urino, M. Mori, T. Nakamura, and Y. Arakawa, "A Hybrid Integrated Light Source on a Silicon Platform Using a Trident Spot-Size Converter," *Journal of Lightwave Technology*, vol. 32, no. 7, 2014.

18. P. Cheben, D.-X. Xu, S. Janz, and A. Densmore, "Subwavelength waveguide grating for mode conversion and light coupling in integrated optics," *Optics Express*, vol. 14, no. 11, 2006.

19. D. Taillaert, F. V. Laere, M. Ayre, W. Bogaerts, D. V. Thourhout, P. Bienstman, and R. Baets, "Grating Couplers for Coupling between Optical Fibers and Nanophotonic Waveguides," *Japanese Journal of Applied Physics*, vol. 45, no. 8A, 2006.

20. R. Polster, L. Y. Dai, O. A. Jimenez, Q. Cheng, M. Lipson, and K. Bergman, "Wafer-scale high-density edge coupling for high throughput testing of silicon photonics," in *Optical Fiber Communication Conference*, Optica Publishing Group, 2018.

21. T. Yoshida, S. Tajima, R. Takei, M. Mori, N. Miura, and Y. Sakakibara, "Vertical silicon waveguide coupler bent by ion implantation," *Optics Express*, vol. 23, no. 23, 2015.

22. M. Trappen, M. Blaicher, P.-I. Dietrich, C. Dankwart, Y. Xu, T. Hoose, M. R. Billah, A. Abbasi, R. Baets, U. Troppenz, M. Theurer, K. Wörhoff, M. Seyfried, W. Freude, and C. Koos, "3D-printed optical probes for wafer-level testing of photonic integrated circuits," *Optics Express*, vol. 28, no. 25, 2020.

23. H. D. Thacker, O. O. Ogunsola, A. V. Muler, and J. D. Meindl, "Wafer-Testing of Optoelectonic Gigascale CMOS Integrated Circuits," *IEEE Journal of Selected Topics in Quantum Electronics*, vol. 17, no. 3, 2011.

24. O. A. J. Gordillo, S. Chaitanya, Y.-C. Chang, U. D. Dave, A. Mohanty, and M. Lipson, "Plug-and-play fiber to waveguide connector," *Optics Express*, vol. 27, no. 15, 2019.

25. X. Leijtens, R. Santos, and K. Williams, "High Density Multi-Channel Passively Aligned Optical Probe for Testing of Photonic Integrated Circuits," *IEEE Photonics Journal*, vol. 13, no. 1, 2021.

26. R. Bates, "The search for the universal probe card solution," in *Proceedings International Test Conference 1997*, Int. Test Conference.

27. T. Gnausch, A. Grundmann, T. Juhasz, T. Kaden, R. Buettner, and T. von Freyhold, "Novel Opto-Electronical Probe Card for Wafer-Level PIC Testing," in *Optical Fiber Communication Conference (OFC) 2019*, OSA, 2019.

28. T. Miura, Y. Maeda, S. Matsuo, and H. Fukuda, "Wafer-level inspection platform on high-volume photonic integrated circuits for drastic reduction of testing time," in *Optical Measurement Systems for Industrial Inspection XI* (P. Lehmann, W. Osten, and A. A. G. Jr., eds.), vol. 11056, International Society for Optics and Photonics, SPIE, 2019.

29. N. C. Harris, Y. Ma, J. Mower, T. Baehr-Jones, D. Englund, M. Hochberg, and C. Galland, "Efficient, compact and low loss thermo-optic phase shifter in silicon," *Optics Express*, vol. 22, no. 9, 2014.

30. M. Milanizadeh, D. Aguiar, A. Melloni, and F. Morichetti, "Canceling Thermal Cross-Talk Effects in Photonic Integrated Circuits," *Journal of Lightwave Technology*, vol. 37, no. 4, 2019.

31. M. Milanizadeh, S. Ahmadi, M. Petrini, D. Aguiar, R. Mazzanti, F. Zanetto, E. Guglielmi, M. Sampietro, F. Morichetti, and A. Melloni, "Control and Calibration Recipes for Photonic Integrated Circuits," *IEEE Journal of Selected Topics in Quantum Electronics*, vol. 26, no. 5, 2020.
32. F. A. Memon, F. Morichetti, and A. Melloni, "High Thermo-Optic Coefficient of Silicon Oxycarbide Photonic Waveguides," *ACS Photonics*, vol. 5, no. 7, 2018.
33. K. Padmaraju and K. Bergman, "Resolving the thermal challenges for silicon microring resonator devices," *Nanophotonics*, vol. 3, no. 4-5, 2014.
34. D. Coenen, H. Oprins, P. D. Heyn, J. V. Campenhout, and I. D. Wolf, "Analysis of Thermal Crosstalk in Photonic Integrated Circuit Using Dynamic Compact Models," *IEEE Transactions on Components, Packaging and Manufacturing Technology*, vol. 12, no. 8, 2022.

Techniques and Methods for Optical Testing

4

4.1 Introduction

After having presented the DUT and the infrastructure to tune it and to test it from the electrical perspective, in this section we introduce a technique to preliminarily calibrate the PIC.

For relatively simple devices, in fact, the measurement of few relevant parameters could be sufficient, but in the presence of reconfigurable, programmable or complex devices, testing is inseparable from control and tuning, which becomes a mandatory precondition to perform quality checks (this is the main difference between PICs and EICs testing [1]).

For these reasons, we propose an effective method capable of:

1. computing spectral deviations between the DUT and a desired REF device, without measuring its spectral response (which is one of the main bottlenecks for fast calibration/testing);
2. classifying DUTs according to specific metrics;
3. tuning the DUT to spectrally resemble (cloning technique) as much as possible the REF;
4. storing actuator working points in a suitable memory structure, LUT, to be used during the actual operative conditions (so that the DUT should be virtually tuned once, during the testing stage).

The final aim of this chapter is to compare these strategies with those already in the literature and to prove that they are compliant, in terms of throughput and resource consumption, with the strict requirements of photonic testing.

© The Author(s), under exclusive license to Springer Nature Switzerland AG 2025   77
M. Petrini, *Mixed-Signal Generic Testing in Photonic Integration*, Synthesis Lectures on Digital Circuits & Systems, https://doi.org/10.1007/978-3-031-60811-7_4

This chapter is organized as follows:

- In Sect. 4.2 an overview of the tuning techniques for complex filters available in the literature is provided.
- In Sect. 4.3 the novel tuning scheme is presented from a theoretical perspective.
- In Sect. 4.4 the novel tuning scheme is validated through suitable numerical simulations.
- In Sect. 4.5 experimental proofs (onto the actual DUT) are presented.
- In Sect. 4.6 the generation of automated LUTs (as optical testing outcomes) is discussed.

4.2   Classical Tuning Techniques

In recent years, many different techniques have been proposed to perform effective and reliable PIC tuning [2–4]. These procedures mainly rely on feedback-controlled systems [5], composed of electronic drivers, actuators and sensors.

Suitable control boards, such as Micro-Processors, Field Programmable Gated Arrays (FPGAs) or custom electronics [6], execute certain algorithms, such as homodyne locking [7], dithering [8, 9] or optimization (i.e., maximum/minimum search) [3]. Their aim is to steer and hold the PIC at the desired working point, by driving the actuators embedded in the considered device. Actuation usually relies on the thermo-optic or the electro-optic effect.

The feedback signals, instead, are provided by proper optical sensors, placed in strategic parts of the circuits, inside the chip or at its output (by means of integrated [10] or transparent photodetectors [11, 12]).

Some of those control algorithms, even if very innovative and effective for high-density programmable circuits, rely on spectral measurements, which are usually avoidable in testing, as we will discuss in the following.

Other approaches, which are typically faster, instead rely only on optical power estimations. These methods have been effectively applied to both very simple structures or to much more complex devices, such as high order MRR filters. In this regard, the results demonstrated by Mak et al. [13] and by Jayatilleka et al. [14] seem to be very interesting, because they are able to align the resonances of several resonators, not relying on any spectral measurements, thus enhancing the convergence speed and reducing the needed instrumentation.

These approaches indeed show the way to follow for effective filter calibration, but they also present some issues that might limit their implementation in a "tuning for testing" scenario.

The main drawback of the latter method, for example, is the small output photocurrent of these photo-resistive-detectors, which leads to a high complexity (and possibly narrow bandwidth) of the electrical read-out circuitry.

Furthermore, both methods exploit narrow-linewidth input signals, whose bandwidth is orders of magnitude smaller than the B_{3dB} of the considered devices. In the former work it is shown that when the process is repeated multiple times on the same device, its responses appear to be quite different (in terms of spectral shape, central wavelength and bandwidth). This is hardly acceptable in a testing scenario.

Indeed, the use of a spectrally broader signal to tune an MRR filter (whose bandwidth matches the bandwidth of the considered PIC) shows better results in terms of spectral repeatability, as in [15]. In this work an effective implementation of the TED, embedded in the control loop is also provided, to make the calibration thermal-cross talk free.

As already stated in the introduction, tuning and testing are deeply connected in photonics, even if they seem to belong to two different worlds.

Before proceeding with further developments, it is worth discussing the typical setup [already introduced in Chap. 2 and here reported again for convenience, in Fig. 4.1a] used to perform optical testing of spectrally selective DUTs. A TLS synchronized with an OSA is exploited to measure the spectral response of the device under test. The deviation between this transfer function and a desired frequency mask is evaluated (through proper metrics such as MSE) by a computational unit, which states if the DUT is suitable or not and how far it is from the desired state. In the presence of a programmable DUT, the computational

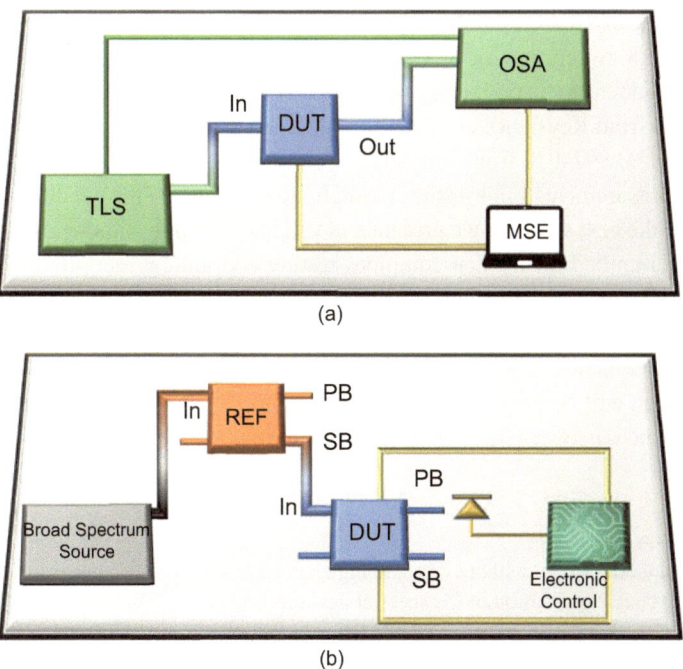

Fig. 4.1 **a** Classical block scheme exploited to tune a filter, composed of the pair TLS–OSA and by a computational unit to calculate MSE. **b** Block scheme exploited in this book, composed of the cascade: BBS–REF–DUT–PD. The electronic board evaluates spectral differences between REF-DUT and keeps under control the DUT. Adapted from [16], with permission

unit could also adjust the PIC actuators, to minimize the said deviation and the spectral measurement is performed again.

It is now convenient to analytically define the absolute MSE, used to evaluate (amplitude) spectral differences:

$$MSE = \frac{1}{N} \sum_{i=1}^{N} (|H_x(\lambda_i)|^2 - |H_y(\lambda_i)|^2)^2 \tag{4.1}$$

where N is the number of acquired samples, while x and y denote the two considered transfer functions.[1]

As already discussed in this book, the quantity in Eq. (4.1) is widely used in many stages of the PICs life-cycle, such as in the design stage (see Sect. 2.3), in the characterization stage (see Sect. 2.4), and in the calibration stage [18].[2]

Even if this approach seems to be common and accurate, it is not suitable for high throughput and high volume testing, since it implies the acquisition of (possibly multiple) spectral response(s), which is time and resource consuming, involving bulky and slow instrumentation.

A typical setup to acquire the amplitude frequency responses (with sufficient accuracy) is, in fact, constituted by the cascade of TLS-DUT-PD or by the cascade of BBS-DUT-OSA or by the cascade of TLS-DUT-OSA. Thus the measurement speed is limited by the slowest between the TLS sweep time or by the OSA scan time. At the time we are writing these lines, top notch devices can perform a wavelength sweep (in the case of TLSs) of a few nm/s (e.g., $N7776C$ from Keysight), or a wavelength scan (in the case of OSAs) in hundreds of samples/s (e.g., $MS9740B$, from Anritsu).[3]

Spectral measurements limit testing throughput, and there is the need to overcome this bottleneck. In the next sections, we present and validate (through numerical simulations and experiments) a unified theory that combines testing and tuning, merging the best of both worlds, i.e., the speed of convergence (and ease electronics) of algorithms based on direct optical power measurements, such as that in [6], and the spectral accuracy of the frequency response measurements.

The approach will be used to perform reliable optical validation (i.e., evaluate possible spectral differences between REF and DUT) and fast "tuning for testing" of a photonic DUT.

---

[1] Use of absolute errors (i.e., without normalizing each addendum of the summation by $|H_x(\lambda_i)|^2$) lead to a more reliable estimation of the spectral deviations [17].

[2] It is worth mentioning that in this work the MSE is calculated between the time domain responses, by acquiring the complex transfer function of the DUT and computing the Inverse Fast Fourier Transform (IFFT).

[3] Spectrometers that can perform frequency scans in millisecond time range are available in the market, such as that in [19]. They could be used in the BBS-DUT-Spectrometer setup. However, their accuracy and repeatability are not proper for complex frequency selective devices, such as those presented in this book.

4.3   Spectral Classification and Cloning Technique

The proposed technique is based on the scheme, reported in Fig. 4.1b.

A BBS, having a constant (and equal to S_0) PSD across the whole operative range of the DUT, is coupled to the input port of the REF. The DUT is then cascaded to the REF. The integral optical power is sensed at the output of the series, by means of a PD.

This book is mainly focused on high-order MRR filters, but this approach is suitable for arbitrary wavelength selective devices (even multiport), having a Passband (PB) and/or Stopband (SB) tuneable spectral responses, $H_{PB}(\lambda)$ and $H_{SB}(\lambda)$, respectively.

We suppose that the DUT has a bandwidth B, and can operate in a wavelength range between λ_1 and λ_2.

Typically, for WDM filters, the passbands show a very flat response, a high out-of-band rejection (at PB port) and a high return loss (at SB port), i.e.:

$$\begin{cases} |\overline{H_{PB}(\lambda)}| >> |\overline{H_{SB}(\lambda)}|, & \text{within } B \\ |\overline{H_{SB}(\lambda)}| >> |\overline{H_{PB}(\lambda)}|, & \text{otherwise} \end{cases} \qquad (4.2)$$

where $\overline{H_{PB}(\lambda)}$ and $\overline{H_{SB}(\lambda)}$ are the nominal responses.

The situation is sketched in Fig. 4.2, where the ideal PB and SB amplitude responses are shown (i.e., $|\overline{H_{PB}(\lambda)}| = 1$ and $|\overline{H_{SB}(\lambda)}| \approx 0$, within B and for lossless devices). PSD coming out from the (passband port of) REF is

$$S_{P,REF}(\lambda) = S_0 |\overline{(H_{PB}(\lambda))}|^2 \qquad (4.3)$$

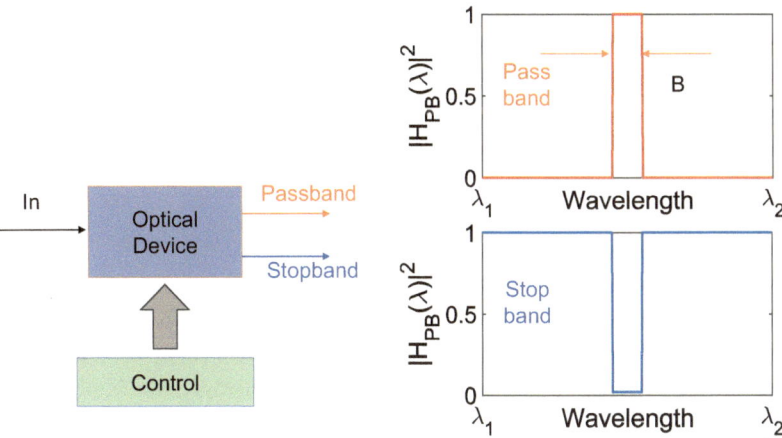

Fig. 4.2 Scheme of a tuneable optical device with two complementary output ports, Stopband and Passband. In-Passband and In-Stopband ideal spectral responses. Adapted from [16], with permission

where the subscript P denotes that the observed port is the PB. This signal is used as source for the DUT, and the coming out PSD (from the passband port) is

$$S_{P,DUT}(\lambda) = S_0 |\overline{(H_{PB}(\lambda))}|^2 |\overline{(H_{PB,DUT}(\lambda))}|^2. \tag{4.4}$$

When the DUT is at the nominal state, i.e. $|\overline{(H_{PB,DUT}(\lambda))}|^2 = |\overline{(H_{PB}(\lambda))}|^2$, the integral output power sensed by the output PD is given by[4]:

$$\begin{aligned}
P_{O_{PP}} &= \int_{\lambda_1}^{\lambda_2} S_{P,DUT}(\lambda)\, d\lambda \\
&= S_0 \int_{\lambda_1}^{\lambda_2} |\overline{(H_{PB}(\lambda))}|^4 (\lambda)\, d\lambda \approx S_0 B.
\end{aligned} \tag{4.5}$$

In these equations losses are not accounted, since they just act as a scaling factor. Furthermore, the last approximation in Eq. (4.5) is valid for a flat passband response, such as that shown in Fig. 4.2. In this scenario, $P_{O_{PP}}$ is the maximum achievable, and the DUT is a spectral clone of the REF.

When the DUT is not in the nominal state (i.e., not properly tuned), its PB response is usually broader, the in-band response is far from flatness (typically it is distorted), resulting in a lower integral output power.

For some devices, such as those in [20], the filter bandwidth is much narrower, and the total output power is lower even in this case. It should be noted that, if the DUT is a tuneable-bandwidth device, the maximum bandwidth of the DUT must be narrower than or equal to the bandwidth of the REF. In fact, if the DUT could assume a bandwidth larger than that of the REF device, the output power maximization procedure described in Eq. 4.5 could lead to an over-sized bandwidth of the DUT with respect to REF.

The same situations also occur in the case of a perfectly tuned DUT, whose central frequency is shifted (by a quantity Δf) with respect to the REF. In this case, the $P_{O_{PP}}$ decreases by a quantity $\Delta P_{O_{PP}} = S_0 \Delta f$, with respect to its maximum ($P_{O_{PP}} = S_0 B$). Thus, the relative output power provides useful information about the central frequency mismatch, as

$$\frac{\Delta P_{O_{PP}}}{P_{O_{PP}}} = \frac{B}{\Delta f}. \tag{4.6}$$

Notably, the sensed optical power could be used as a figure of merit to estimate the similarity between REF and DUT. In addition, it can be used as a feedback signal for the implementation of the calibration techniques, to make the DUT response ($|H_{PB,DUT}(\lambda)|$) resemble the REF frequency behaviour (cloning). In particular, the quantity of $P_{O_{PP}}$ should be maximized.

Similar considerations also hold in the case of SB ports, for both DUT and REF. Even in this case the tuning criterion is the maximization of the feedback signal (namely, the sensed optical power), which in principle guarantees spectral cloning. However, in practice, device

---

[4] Power coming out from the cascade REF (Passband)-DUT (Passband).

comparison mainly relies on their out-of-band behaviour and the results are less accurate, with respect to the PB situation.

Hybrid conditions, such as REF PB and DUT SB (or viceversa), can be used as well, but they require the minimization of the output power, which is given by[5]:

$$
\begin{aligned}
P_{O_{PS}} &= S_0 \int_{\lambda_1}^{\lambda_2} |\overline{H_{PB}(\lambda)}|^2 |H_{SB,DUT}(\lambda)|^2 \, d\lambda \\
&= S_0 \int_{\lambda_1}^{\lambda_2} |\overline{H_{PB}(\lambda)}|^2 (1 - |H_{PB,DUT}(\lambda)|^2) \, d\lambda \\
&\approx S_0(B - (B - \Delta f)) = S_0 \Delta f,
\end{aligned}
\tag{4.7}
$$

where the approximation holds for a flat PB response, frequency shifted by a quantity Δf, with respect to the target REF device.

The absolute power variation is $P_{O_{PS}} = S_0 \Delta f$, and $P_{O_{PS}} \approx 0$ in the case of perfect tuning, considering again the losses as just a scaling factor.

The two combinations, PB REF–PB DUT and PB REF–SB DUT, are shown in Fig. 4.3(a) and Fig. 4.3(c), respectively. A qualitative dependence of the sensed output power against the (REF–DUT) detuning and the spectral misalignment, are shown in Fig. 4.3b, d, for the two accounted scenarios.

In general, the combination of two devices, having complementary behaviour (i.e., PB and SB), shows a better sensitivity for DUT tuning, with the minor disadvantage of working with relatively low optical power levels.

Hence, in the next section we demonstrate, through suitable numerical simulations, the correlation linking the MSE (between the spectral responses of two frequency-selective devices) and the two aforementioned quantities, $P_{O_{PP}}$ and $P_{O_{PS}}$.

These values can be advantageously used not only for real time PIC control, but also for calibration, testing and classification purposes.

The method is, in principle, valid for arbitrary circuit topologies (having different control variables and actuators),[6] implemented in different platforms, but with frequency selective features.

Before concluding the Section, a couple of remarks have to be addressed.

---

[5] Power coming out from the cascade REF (Passband)-DUT (Stopband).

[6] The method is valid even when one output port is not accessible. Furthermore it can be extended to the multiport case (i.e., devices with N output ports), such as Arrayed Waveguide Gratings (AWG), satisfying the same conditions given in this Section. In the case of power maximization out of the cascade REF (passband)-DUT (passband), the approach works by considering the j-th DUT passband port, $|H_{PB}(\lambda)|^2 = |H_j(\lambda)|^2$. Instead, in the case of power minimization out of the cascade REF (passband)-DUT (stopband), we need to minimize the overall power $P_{O_{PS}}$ at the dark ports of the DUT, $P_{O_{PS}} = S_0 \int_{\lambda_1}^{\lambda_2} |\overline{H_{PB}(\lambda)}|^2 T_{DUT}(\lambda) \, d\lambda$, where $T_{DUT}(\lambda) = \sum_{i=1, i \neq j}^{N} |H_i(\lambda)|^2$ and $|H_i(\lambda)|^2$ is the amplitude spectral response of the i-th DUT output port.

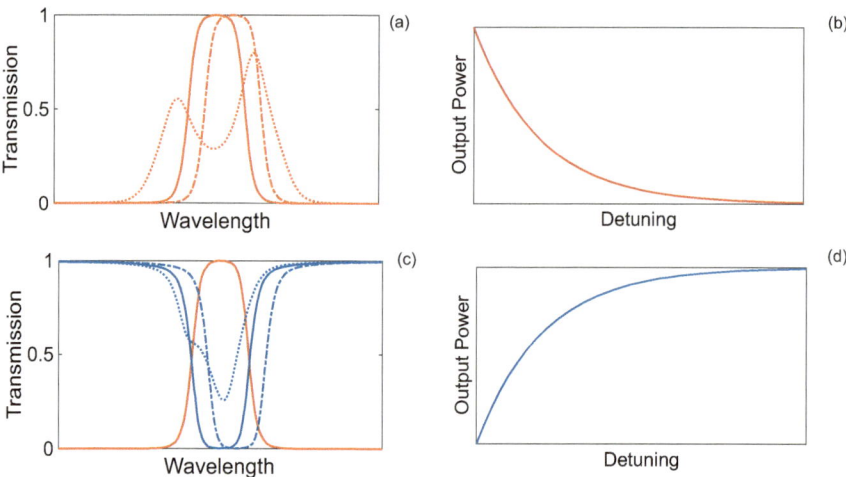

Fig. 4.3 **a** REF ideal passband spectral response (solid orange line) and DUT perturbed passband responses. **b** $P_{O_{PP}}$ vs detuning or shifting. It is a maximum when the REF and DUT responses match. **c** Same of (**a**) for passband-stopband combination. **d** $P_{O_{PS}}$ versus detuning or shifting. It is a minimum when the REF and DUT responses match. Adapted from [16], with permission

First, the REF and the DUT do not necessarily have to be integrated in the same chip (or made in the same technology/topology). In the case of wafer-level-testing, for example, an external REF can be cascaded to a broadband source and placed before the optical probing station, coupling into the DUT an optical beam with a suitably shaped PSD.

Conveniently, the external REF can be a packaged device with its control electronics, which has been previously calibrated and well stabilized against thermal drifts.

Second, this testing method is based on an intensity monitoring, thus in general it does not allow for tuning of the DUT phase response. However, in the case of all-pole devices (such as the In-Drop response of the filter described in Chap. 2) the imaginary part of the spectral is a bijective function of the real part [21]. Hence, once the optimum is reached, the phase dispersion can be recovered without ambiguity (net of a constant offset related to the geometrical length of the photonic device).

The approach could also be employed together with dimensionality reduction and machine learning strategies [22].

4.4   Numerical Simulations

In this section we present some numerical simulations to prove that $P_{O_{PP}}$ and $P_{O_{PS}}$ could be effectively exploited as a reliable estimator of the MSE between a REF and a DUT.

First, we consider the situation sketched in Fig. 4.4a, where REF[7] and DUT are ideal MZIs, with perfect 50:50 directional couplers. Their transfer function (in power) is $|H_{MZ,C}(\omega)|^2 = \cos^2(\omega T)$ for the cross port (denoted by the subscript C) and $|H_{MZ,B}(\omega)|^2 = \sin^2(\omega T)$ for the bar port (denoted by the subscript B). ω is the angular frequency and T is the imbalance of the two MZIs, in time units.

Let us now suppose that the BBS coupled to the input of the REF device is broader (in terms of spectral width) than a single FSR of the MZIs, namely $FSR = 1/T$ [Hz]. In addition, the two interferometers are frequency misaligned by a quantity $\Delta\omega$. The spectral responses of the REF (solid red curve) and DUT (dashed red curve) and their product (dashed blue curve) are shown in Fig. 4.4b.

The integral output power can be analytically computed as follows

$$P_O = S_0 \int_{-FSR/2}^{FSR/2} |H_{REF,C,B}(\lambda)|^2 |H_{DUT,C,B}(\lambda)|^2 \, d\lambda$$

$$\propto \pm\cos(\Delta\omega T) \tag{4.8}$$

where for both the REF and the DUT we can arbitrarily choose cross or bar ports. The $\pm$ sign depends on the devices' relative configuration, i.e., if the two MZIs are connected to the same port ($-$) or crossed ($+$).

According to Eq. (4.1) (or better, according to its continuous domain counterpart), the MSE could also be analytically computed (clearly between the same port of the two devices, i.e. bar-bar or cross-cross) and it is

$$MSE \propto -\cos(\Delta\omega T). \tag{4.9}$$

Figure 4.4c shows the MSE and the output powers, computed in different situations, on the same plot. The curves have been normalized and properly offset, in such a way that the maxima are equal to 1 and the minima to 0. In particular $P_{O_{CB}}$ refers to the cascade REF (cross)–DUT (bar), while $P_{O_{CC}}$ refers to the cascade REF (cross)–DUT (cross).

This plot shows that the integral optical power sensed at the output of the cascade provides direct information for the evaluation and possibly the minimization of the MSE. Notably, the absolute values (at least in simulations) do not affect the validity of the statement.

Similar speculations can be carried out considering other simple devices, such as MRRs, Fabry-Pérot cavities and Bragg gratings. In principle, the more selective is the device in frequency, the wider the dynamic range of the sensed output power.

---

[7] From now on REF and DUT are identical from a circuit point of view. However, the technique is effective even if it differs from topological and technological standpoints.

Fig. 4.4 a Cascade of MZI–REF and MZI–DUT for testing and calibration purposes. **b** Spectral response of REF (red solid line), DUT (red dashed line) and their product (blue dashed line). **c** Dependence of the MSE and the P_O against $\Delta\omega$. The normalized P_O sign depends on the REF–DUT connection (i.e., P_{OCB} or P_{OCC}). P_{OCB} in the case of REF (cross)–DUT (bar) (or viceversa) and P_{OCC} in the case of REF (cross)–DUT (cross). Adapted from [16], with permission

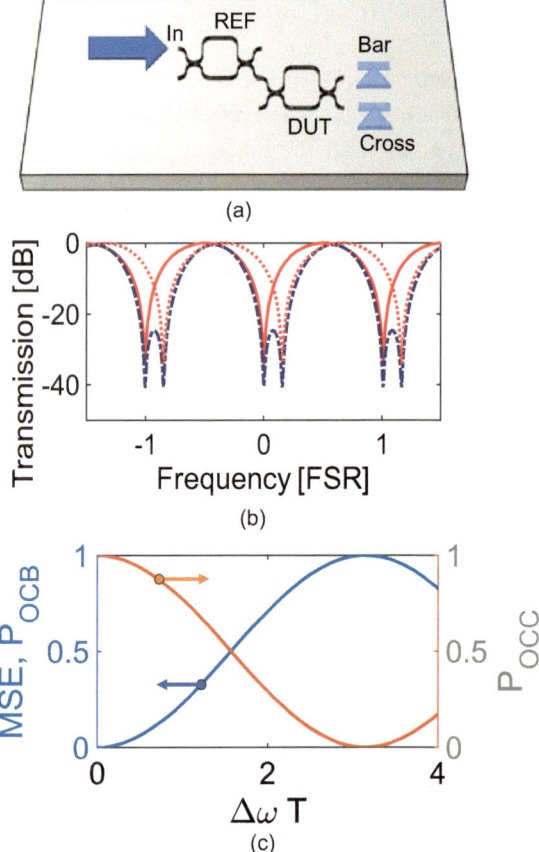

As a second example, we consider a more complex PIC, similar to that introduced in Chap. 2.[8] The device, sketched in Fig. 4.5a, is an MRR-based filter. It is characterized by a bandwidth $B = 40$ GHz, MRRs radii $r = 10\ \mu$m, and a group index $n_g = 3.89\ (FSR = 1.2$ THz), with power coupling ratios equal to [0.282, 0.0099, 0.0035, 0.0085, 0.236]. We recall that those coupling ratios provide an optimized Chebyshev response and, that they are slightly asymmetric, because the round-trip losses have been accounted for ($\alpha = 0.02$ dB/turn).

The REF Drop port (PB) is connected to the input of the DUT filter, and the output power can be sensed either at its Drop ($P_{O_{PP}}$) or at its Through ($P_{O_{PS}}$).

The nominal spectral response of this kind of device (REF In-Drop transfer function, $\overline{|H_{PB}(\lambda)|^2}$) is shown in Fig. 4.5b (red solid line), as well as the REF-DUT cascade in nominal condition, $\overline{|H_{PB}(\lambda)|^4}$ (red dashed line).

---

[8] It is the non-Vernier counterpart of the filter designed in Chap. 2.

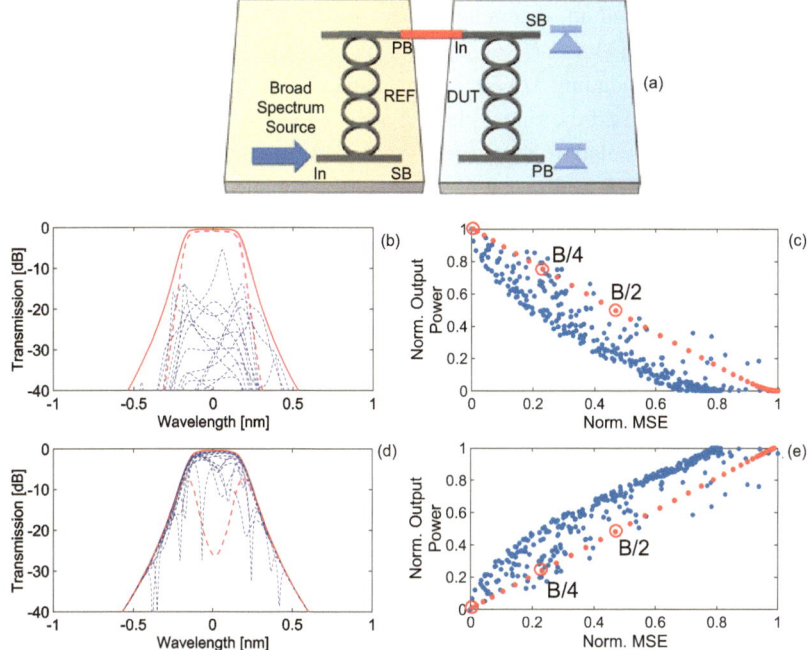

Fig. 4.5 a Cascade of filter-REF and filter-DUT for testing and calibration purposes. **b** Spectral response of Passband-REF (red solid line). Spectral response of the cascade Passband-REF—Passband-DUT, in case of DUT perfectly tuned (red dashed line) and in the case of perturbed DUT (blue dashed lines). **c** Collected output power against the MSE between the two In-Passband spectral responses. Their correlation is evident, and the power is maximized when MSE is ≈0. The red dots represent DUT nominally tuned, but rigidly shifted in the frequency domain. **d–e** Same as (**b**)–(**c**) considering Stopband port of the DUT. The output power is minimal when MSE is ≈0. For the sake of visualization not all the simulated cases are reported in (**b**)–(**d**). Adapted from [16], with permission

To resemble a realistic case, uncorrelated phase perturbations, picked up from a uniform distribution, within the interval $\pm\pi$,[9] are applied to each MRR of the filter, while its coupling ratios are assumed to be equal to the nominal value.

The spectral responses at the Drop output of the DUT, $|\overline{(H_{PB}(\lambda))}||H_{PB,DUT}(\lambda)|^2$ are reported as blue dashed lines in Fig. 4.5b. The total MSE and the integral output power, i.e. $P_{O_{PP}}$ (under the assumption that the input BBS has a PSD S_0 broader than B), have been computed for each simulated case.

The total MSE is conveniently defined as

$$MSE_{TOT} = MSE_{SB} + MSE_{PB}, \tag{4.10}$$

---

[9] One hundred different randomly perturbed DUTs are accounted for.

where the single addendum is computed according to Eq. (4.1), between the pairs $H_{DUT,SB}(\lambda)-H_{REF,SB}(\lambda)$ and $H_{DUT,PB}(\lambda) - H_{REF,SB}(\lambda)$, respectively.

$P_{O_{PP}}$ and the corresponding MSE_{TOT} (normalized to its maximum, i.e. when a perturbation of π is applied to all the rings' phases) are reported in the scatter plot, in Fig. 4.5c. There is an evident correlation between the two, and, in particular, the output power is maximized when the MSE is almost null and it accordingly decreases.

Conversely, in the case of observation of the output power at the Through port of the DUT, the output power vanishes ($P_{O_{PS}}$) when the two devices spectrally resemble each other (i.e., $MSE \approx 0$), as reported in the scatter plot $P_{O_{PS}}$ vs MSE_{TOT} (again, normalized to its maximum), in Fig. 4.5e. The PSD at the output of the cascade (for each considered case), $|\overline{(H_{SB}(\lambda))}|^2|H_{PB,DUT}(\lambda)|^2$ is shown in Fig. 4.5d, as dashed blue line. The case regarding DUT in perfect conditions is reported in the same picture as red dashed line.

Furthermore, in Fig. 4.5c, e, the red dots represent, respectively, the behavior of $MSE - P_{O_{PP}}$ and $MSE - P_{O_{PS}}$ is the case of a DUT nominally tuned, shifted by Δf, with respect to the REF. As expected from previous considerations, $P_{O_{PP}}$ linearly decreases with Δf, becoming null when $\Delta f > B$ (opposite behaviour is shown for $MSE - P_{O_{PS}}$, and $P_{O_{PS}}$ approaches 0, for $\Delta f = 0$).

Therefore, in a testing scenario, it is possible to analyse the $P_{O_{PS}}$ and/or the $P_{O_{PP}}$, in place of measuring spectral responses and computing the MSEs. Putting thresholds on the optical power values enables the classification and clustering of the DUTs.

Advantageously, during real experiments and real testing procedures, the ratio $P_{O_{PP}}/P_{O_{PS}}$ could be effectively evaluated, to remove uncertainties in the absolute power, that may arise due to fiber-chip coupling tolerances or due to oscillations of the BBS.

4.5   Experimental Measurements

In this paragraph, the experimental evidence of the approach presented in Sects. 4.3 and 4.4 is provided. The actual device under test is the 4-channel-TOADM, widely described in Chap. 2.

According to the method we want to implement, to perform the calibration and optical validation we need a BBS and a reference device. A scheme of the experimental setup is shown in Fig. 4.6a.

As broadband source we used the Amplified Spontaneous Emission (ASE) noise, from an EDFA, but a superluminescent diode or other sources could be also used as well.

As a REF device we exploit an external filter (single polarization, topologically and technologically identical to the single channel of the TOADM), tuned to its own nominal condition, according to the procedure outlined in Fig. 4.1a. Its transfer functions (for both Through and Drop ports) are reported again in Fig. 4.6b, for readers' convenience. In particular the Drop response is also the PSD employed to test the TOADM itself.

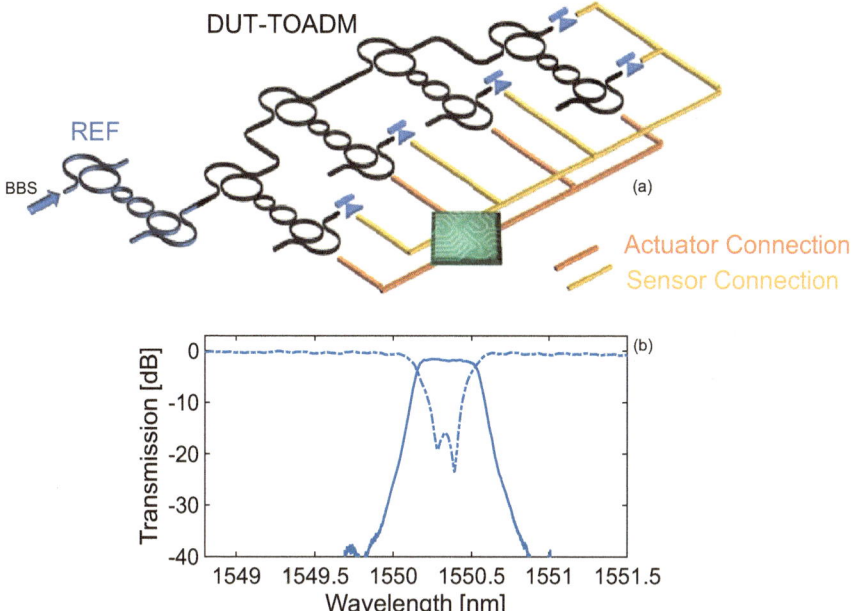

Fig. 4.6 a Setup exploited to test and tune the presented TOADM. The single channel REF filter is highlighted in blue. **b** Its spectral responses (Through port, dashed line and Drop port, solid line)

As already discussed, each channel of the TOADM under test could be effectively disconnected by the combined use of integrated VOAs and TCs. This enables sequential tuning and testing of the different filters of the multiplexer, since only one of them at a time should be connected (and tested), while all the others are disconnected.

Before the actual tuning, the natural spectral responses of the filters have a random behaviour, due to the spread of MRR resonances, caused by fabrication imperfections. These non-testable spectral responses are shown in Fig. 4.7a (for Through port) and 4.7b (for Drop port), where we considered eight different channels (belonging to two different TOADMs).

The electronic control architecture, by using the information of the PDs placed at the Drop ports of the filters and of an external PD coupled to the Through port, drives the thermo-optic actuators of the DUT according to a steepest descent algorithm [15].

In particular, we perform two steps:

1. $P_{O_{PP}}$[10] maximization, to achieve coarse but fast tuning in the case of spectral response is far from the target;

---

[10] Power coming out from the cascade REF (Drop)–DUT (Drop).

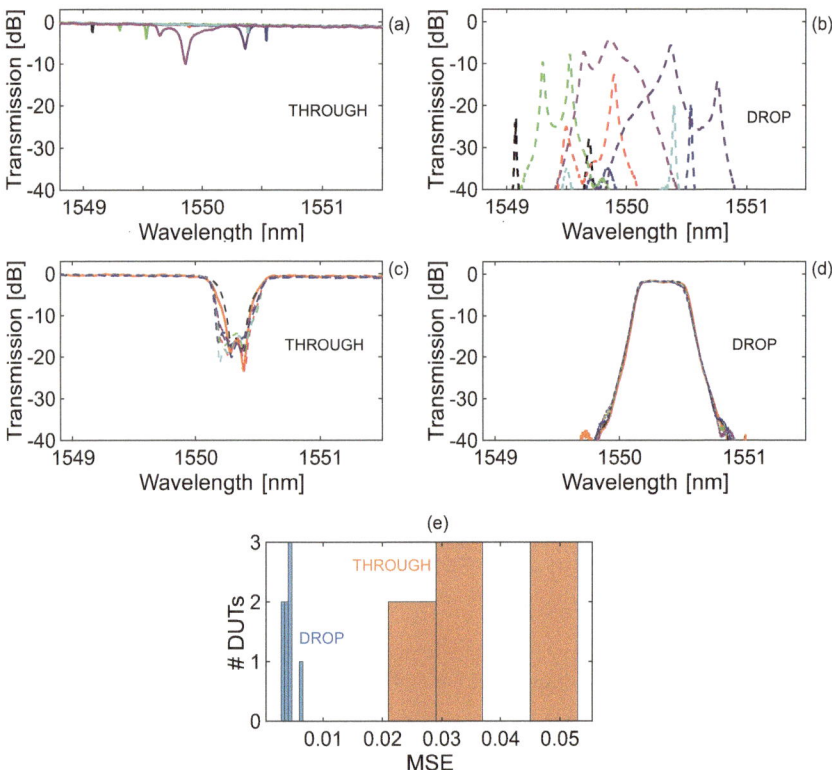

Fig. 4.7 Natural responses of the TOADM channels for Through (**a**) and Drop (**b**) ports. Tuned (by using the proposed method) responses of the TOADM channels for Through (**c**) and Drop (**d**) ports. Orange solid lines represent the REF transfer functions. (**e**) MSE between REF and each channel, for the cases In-Drop (in blue) and In-Through (in orange). The latter is one order of magnitude greater than the former. Adapted from [16], with permission

2. $P_{O_{PS}}$[11] minimization, to achieve precise and accurate tuning.

The acquired spectra of all eight tuned channels were acquired, choosing a wavelength span of 2.6 nm and a wavelength sampling of 1 pm. The results are reported in Fig. 4.7c, d, for both Through and Drop ports, respectively. In these pictures orange solid lines represent the REF transfer functions. For all the considered cases, an excellent spectral overlap can be observed.

The main features of the employed control loop are as follows:

---

[11] Power coming out from the cascade REF (Drop)–DUT (Through).

1. the applied voltage accuracy to each actuator, $\Delta V = 0.5\,\mathrm{mV}$ (resolution of DAC cascaded with its own driver is $0.3\,\mathrm{mV}$). This is suitable for this kind of PIC;
2. the heater control signals are weighted by specific coefficients that come from the TED technique (to make the calibration thermal crosstalk free, see Chap. 3). In this regard, the importance of matrix **T**, computed during the electrical validation, becomes crucial;
3. the heater control signals are progressively adapted to optimize the speed and the accuracy of the convergence (i.e., the driving voltage steps progressively decrease as the sensed optical power is optimized).

Notably, the DUT and REF substrates were thermally stabilized by a closed-loop TEC, and kept at 25 °C.

For all the considered cases, the control algorithm converged in <100 ms, or few tens of iterations (less than the fiber-chip coupling routine duration, as discussed in Chaps. 1 and 3). Notably, before executing the (i+1)-th iteration, we have to wait the end of the transients of the i-th iteration, not only related to the thermo-optic time constant (in the order of μs time range), but is also related to the thermal crosstalk time constant (in the order of ms time range, see Sect. 3.6). Thus, the latter is the actual lower bound for the convergence speed of the algorithm.

Finally, the MSE between the reference and each DUT was evaluated, and reported in the histogram in Fig. 4.7, for both output responses. It can be observed that the MSE referred to the Through port is approximately one order of magnitude greater than that referred to the Drop port.[12] This highlights the fact that the minimization of $P_{O_{PS}}$ gives more accurate metrics for filter tuning, than the maximization of $P_{O_{PP}}$.

From a quantitative standpoint, the goodness of this "tuning-for-testing", can be evaluated by measuring some relevant parameters and computing their standard deviation. Specifically, in Fig. 4.8 shows three histogram plots, related to insertion loss (a), B_{3dB} (b) and return loss (c). Not only all these values are close to those of the REF filter (and so we can conclude that all the channels of the TOADM passed the test), but their standard deviation are also quite small. In fact, variability is <0.12 dB for insertion loss, <0.60 GHz for B_{3dB} and < 2 dB for return loss (averaged over 20 GHz around the central frequency).

To conclude, for us "testing" means stating whether a DUT (i.e. filter) is able (or not) to reach the spectral status of the REF, which represents the desired specifications. To do so, we do not rely on frequency-domain measurements but we measure (and optimize, since the DUT is programmable) the power coming out from the cascade REF-DUT, proving that it is a good estimator of REF-DUT spectral deviation (MSE). Thus, setting a threshold on this amount of power is equivalent to setting a threshold on the MSE. The optimum threshold comes from the combination of the preliminary measurements/simulations and the testing

---

[12] These values of MSE have been obtained by stopping the control loop around the optimum point, with the residual optical power oscillations <0.05 dB, well detectable by the embedded sensors and the proposed signal acquisition chain (corresponding to an output voltage oscillation of the TIA $<10\,\mathrm{mV}$). If convergence cannot be reached in 150 iterations the test is considered to have failed.

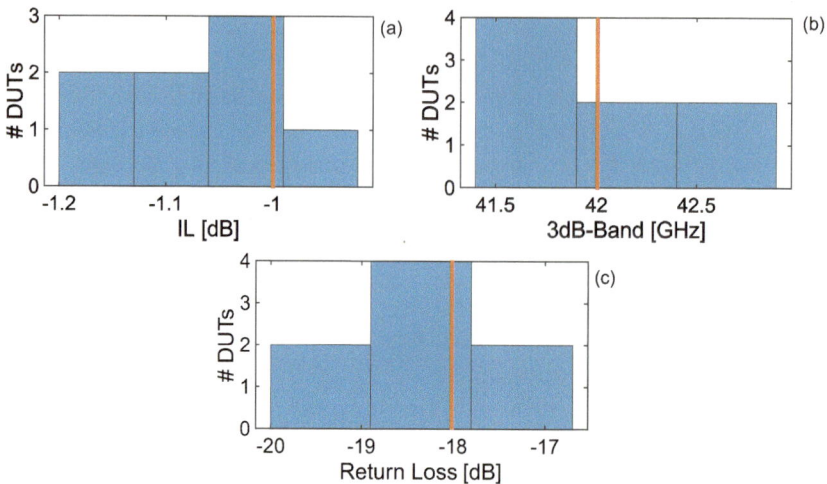

Fig. 4.8 Histograms representing Insertion Loss (**a**), $3dB$-bandwidth (**b**) and Return Loss (**c**) of the tuned filters by using the proposed approach. The orange lines represent nominal values. Adapted from [16], with permission

accuracy/robustness. For example, we have chosen values of normalized (over the input intensity) output power of $P_{O_{PP}}/P_{IN} > 0.8$ or $P_{O_{PS}}/P_{IN} < 0.2$. These values turn into an MSE <0.1 and good accordance with the target features.

However, regarding the phase response, with this method we cannot guarantee that the device is in line with specs. Since the DUT passband response is all-pole, the phase response is a bijective function of the amplitude response (net of a constant offset). In general this does not hold. Nevertheless, phase observation requires time-consuming interferometric techniques, that are not currently suitable for volume testing and will be the subject of future investigations.

The reliability, speed and accuracy of this method strongly depend on the characteristics of the employed sensors, actuators and controllers. For this specific aim (the calibration of a photonic filter bank), the features summarized in Chap. 3 are suitable (for example in terms of the detectable power dynamic range, the accuracy of heater control, and the accuracy of current read). However, it must be highlighted that the testing system must be carefully tailored to the specifications of the DUT.

4.6   Automated LookUp Table Generation

The LUTs are very well established in information technology, and successfully exploited, for example, in digital electronics. These tables contain a priori determined values (which

Table 4.1 LUT for the four-channels-TOADM under test. Its content is the electrical power consumption of the heaters controlling the building blocks of the whole PIC

Building block	Channel 1 [mW]	Channel 2 [mW]	Channel 3 [mW]	Channel 4 [mW]
Top MZI	4.16	3.91	14.64	16.26
Ring 1	11.98	13.27	14.26	14.47
Ring 2	7.96	8.84	10.17	9.93
Ring 3	4.58	5.35	5.14	4.59
Ring 4	8.08	9.10	9.69	9.36
Bottom MZI	11.08	10.04	12.07	11.58

completely describe the working condition of a system), and by using them the state of a certain device could be easily recovered, avoiding the execution of other calibration loops.

Usually, building an LUT for a PIC could be very time consuming (during testing or characterization stages), since its initial (natural) condition may be far away (and possibly unknown) from the target, as discussed in the previous sections.

The control methods presented so far, allowing for a precise, fast and automatic tuning of the DUT, into a specific working point, could be advantageously used to fill a LUT.

Thus, in this scenario, once the cloning technique reaches convergence and the single DUT is validated from both the optical and electrical standpoints, its actuators' status (namely, the driving voltage, the driving current or the absorbed electrical power) can be stored in proper memory structures.

Since the DUT is a reconfigurable multiplexer, the central wavelength of the REF device could be moved along the operational wavelength range (according, for example, to the ITU-T grid) and the cloning technique could be executed, multiple times, one per wavelength channel. In so doing, we obtain a set of working points, that fully describe the shape of the whole filter bank. The spectral responses (Through and Drop ports) of the completely tuned TOADM and the relative LUT (in the electrical power consumption domain) are reported in Fig. 4.9 and in Table 4.1. Each filter is tuned to its nominal condition (by using the cloning technique), following the 200-GHz- ITU-T grid (channel 1 at 1536.61 nm, channel 2 at 1538.19 nm, channel 3 at 1539.79 nm and channel 4 at 1541.37 nm).

By using an active and real-time control loop, the LUT could be updated during the normal operation of the PIC, in the case of sudden changes of the working conditions, such as temperature shift in the environment, or in the case of different spectral shapes of the input signals.[13] As demonstrated in [15] the spectral response of a high-order-MRR-based filter could be matched onto the PSD of the input signal, according to certain figures of merit.

---

[13] In flexible networks, with dynamic allocation of the bandwidth, different signals are transmitted, having different constellations, modulation formats, bit-rates and pulse shapes.

In conclusion, we believe that LUTs are becoming a key element for novel programmable photonic integrated architectures and should be an outcome of the testing stage. An LUT, in fact, increases the value of a PIC and makes its exploitation simpler for the final user.

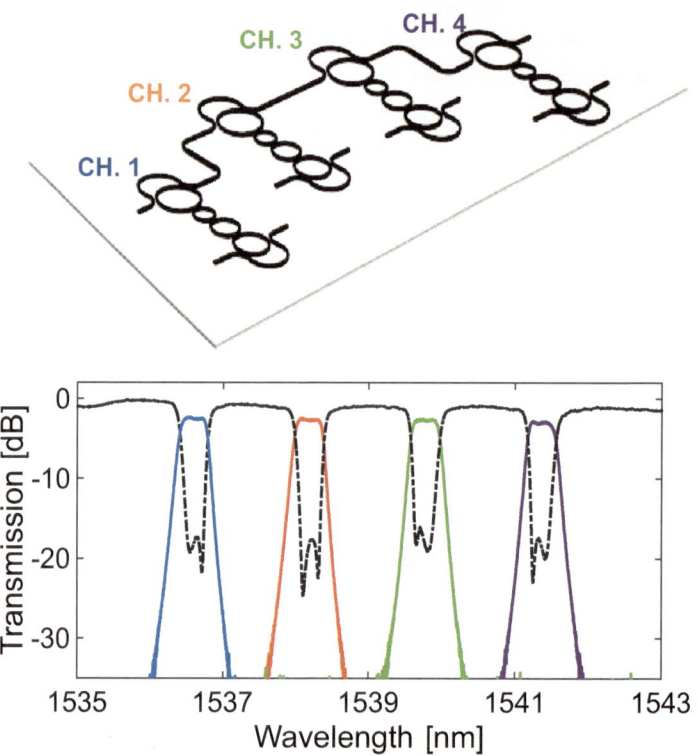

Fig. 4.9 TOADM with all channels simultaneously tuned, according to the standard ITU-T grid. The working points of its actuators are stored in a proper LUT

References

1. V. A. M. Bushnell, *Essentials of Electronic Testing for Digital, Memory and Mixed-Signal VLSI Circuits.* Springer US, 2006.
2. A. Annoni, E. Guglielmi, M. Carminati, G. Ferrari, M. Sampietro, D. A. Miller, A. Melloni, and F. Morichetti, "Unscrambling light—automatically undoing strong mixing between modes," *Light: Science & Applications*, vol. 6, no. 12, 2017.
3. J. C. C. Mak and J. K. S. Poon, "Multivariable Tuning Control of Photonic Integrated Circuits," *Journal of Lightwave Technology*, vol. 35, no. 9, 2017.

4. H. Jayatilleka, K. Murray, M. Á. Guillén-Torres, M. Caverley, R. Hu, N. A. F. Jaeger, L. Chrostowski, and S. Shekhar, "Wavelength tuning and stabilization of microring-based filters using silicon in-resonator photoconductive heaters," *Optics Express*, vol. 23, no. 19, 2015.

5. F. Morichetti, S. Grillanda, and A. Melloni, "Breakthroughs in Photonics 2013: Toward Feedback-Controlled Integrated Photonics," *IEEE Photonics Journal*, vol. 6, no. 2, 2014.

6. Y. Li and A. W. Poon, "Active resonance wavelength stabilization for silicon microring resonators with an in-resonator defect-state-absorption-based photodetector," *Optics Express*, vol. 23, no. 1, 2015.

7. J. A. Cox, A. L. Lentine, D. C. Trotter, and A. L. Starbuck, "Control of integrated micro-resonator wavelength via balanced homodyne locking," *Optics Express*, vol. 22, no. 9, 2014.

8. K. Padmaraju, D. F. Logan, T. Shiraishi, J. J. Ackert, A. P. Knights, and K. Bergman, "Wavelength Locking and Thermally Stabilizing Microring Resonators Using Dithering Signals," *Journal of Lightwave Technology*, vol. 32, no. 3, 2014.

9. F. Zanetto, V. Grimaldi, F. Toso, E. Guglielmi, M. Milanizadeh, D. Aguiar, M. Moralis-Pegios, S. Pitris, T. Alexoudi, F. Morichetti, A. Melloni, G. Ferrari, and M. Sampietro, "Dithering-based real-time control of cascaded silicon photonic devices by means of non-invasive detectors," *IET Optoelectronics*, vol. 15, no. 2, 2021.

10. P. Dumais, D. J. Goodwill, D. Celo, J. Jiang, C. Zhang, F. Zhao, X. Tu, C. Zhang, S. Yan, J. He, M. Li, W. Liu, Y. Wei, D. Geng, H. Mehrvar, and E. Bernier, "Silicon Photonic Switch Subsystem With 900 Monolithically Integrated Calibration Photodiodes and 64-Fiber Package," *Journal of Lightwave Technology*, vol. 36, no. 2, 2018.

11. S. Grillanda, M. Carminati, F. Morichetti, P. Ciccarella, A. Annoni, G. Ferrari, M. Strain, M. Sorel, M. Sampietro, and A. Melloni, "Non-invasive monitoring and control in silicon photonics using CMOS integrated electronics," *Optica*, vol. 1, no. 3, 2014.

12. F. Morichetti, S. Grillanda, M. Carminati, G. Ferrari, M. Sampietro, M. J. Strain, M. Sorel, and A. Melloni, "Non-Invasive On-Chip Light Observation by Contactless Waveguide Conductivity Monitoring," *IEEE Journal of Selected Topics in Quantum Electronics*, vol. 20, no. 4, 2014.

13. J. C. C. Mak, W. D. Sacher, T. Xue, J. C. Mikkelsen, Z. Yong, and J. K. S. Poon, "Automatic Resonance Alignment of High-Order Microring Filters," *IEEE Journal of Quantum Electronics*, vol. 51, no. 11, 2015.

14. H. Jayatilleka, H. Shoman, L. Chrostowski, and S. Shekhar, "Photoconductive heaters enable control of large-scale silicon photonic ring resonator circuits," *Optica*, vol. 6, no. 1, 2019.

15. M. Milanizadeh, S. Ahmadi, M. Petrini, D. Aguiar, R. Mazzanti, F. Zanetto, E. Guglielmi, M. Sampietro, F. Morichetti, and A. Melloni, "Control and Calibration Recipes for Photonic Integrated Circuits," *IEEE Journal of Selected Topics in Quantum Electronics*, vol. 26, no. 5, 2020.

16. M. Petrini, M. Seyfried, F. Morichetti, and A. Melloni, "Spectral Classification and Cloning of Photonic Integrated Filters for Volume Testing," *Journal of Lightwave Technology*, vol. 41, no. 1, 2023.

17. U. Spagnolini, *Statistical Signal Processing*. 2017.

18. X. Xu, G. Ren, T. Feleppa, X. Liu, A. Boes, A. Mitchell, and A. J. Lowery, "Self-calibrating programmable photonic integrated circuits," *Nature Photonics*, vol. 16, no. 8, 2022.

19. A. H. Martin Finger, Frank Muenchow, "Testing Solutions for VCSELs With High Resolution Array Spectroradiometer (White Paper)," *Instrument Systems*, 2019.

20. P. Orlandi, F. Morichetti, M. J. Strain, M. Sorel, P. Bassi, and A. Melloni, "Photonic Integrated Filter With Widely Tunable Bandwidth," *Journal of Lightwave Technology*, vol. 32, no. 5, 2014.

21. A. Mencozzi, "A necessary and sufficient condition for minimum phase and implications for phase retrieval," *Trans. Inf. Theory*, vol. 13, no. 9, 2014.

22. W. Gao, L. Lu, L. Zhou, and J. Chen, "Automatic calibration of silicon ring-based optical switch powered by machine learning," *Optics Express*, vol. 28, no. 7, 2020.

Mitigation of Nonlinear Effects

5

5.1 Introduction

In certain nodes of real networks high power signals may be present and nonlinear effects limit the maximum optical intensity that can be handled by photonic integrated circuits. This limitation is stricter in silicon photonics, working at typical telecommunication wavelengths (usually <10 dBm, approximately 1550 nm). Even if the nonlinearities are usually advantageously exploited for frequency comb generation [1] and frequency conversion [2], they are detrimental for linear applications, such as filtering and multiplexing.

Moreover, these effects have an even worse impact on MRR-based devices (and more in general on resonant circuits), such as the considered DUT, due to the intracavity field enhancement. For these reasons, after discussing the physical phenomenology of the nonlinearities and a few approaches implemented to address them, we investigate their impact directly on the considered DUT, by means of appropriate (device-level and system-level) measurements. We also propose the use of the control system described in Chap. 4 to mitigate the nonlinear effects and to effectively tune the filter in the presence of high-power signals. In so doing we can extend the validity range of the testing technique.

Furthermore, we discuss how the LUT can be updated to make the DUT ready for high-intensity inputs (i.e., predistortion of the filter).

This chapter is organized as follows:

- In Sect. 5.2 typical nonlinear effects are summarized and some common practices to counteract them are listed.
- In Sect. 5.3 the impact of nonlinear effects on the TOADM is assessed and the benefit of the proposed method is stated.
- In Sect. 5.4 we describe the LUT update, to obtain the DUT predistortion.

© The Author(s), under exclusive license to Springer Nature Switzerland AG 2025　　　　97
M. Petrini, *Mixed-Signal Generic Testing in Photonic Integration*, Synthesis Lectures on Digital Circuits & Systems, https://doi.org/10.1007/978-3-031-60811-7_5

5.2   Nonlinear Effects in Silicon Photonics

5.2.1   Nonlinear Phenomena

Silicon PICs present a number of nonlinear effects in the near-IR wavelength range (1300 − 1700 nm) [3, 4]. These are triggered at a relatively low optical power threshold, typically above a few tens of mW, due to the combination of three elements:

- low energy gap (approximately 1.1 eV in the near-IR spectral window),
- large $\chi^{(3)}$ coefficient, or equivalently large $n_2 = 4.5 \cdot 10^{-18}$ m^2/W (Kerr coefficient),
- small waveguide effective area (A_{eff}). Since the optical mode is well confined (as it usually occurs in high index contrast platforms), A_{eff} could be considered approximately equal to the geometrical area, in our case $A_{geom} = (500\ \text{nm})(220\ \text{nm}) \approx 0.1\ \mu\text{m}^2$.

In such a small bandgap material, when the optical input power is sufficiently high, the photons are absorbed [Two Photon Absorption (TPA)] and free carriers are generated (in the conduction band). As already discussed in Chap. 2, when there are free carriers in the waveguide core, the real part (due to the FCD) and imaginary part (due to the FCA) of the refractive index change, proportionally to the number of those free carriers (which is proportional to the incoming optical power). From a practical point of view, this means that the effective refractive index (n_{eff}) decreases [Eq. (2.4)] and, at the same time, the attenuation coefficient increases [Eq. (2.3)], by certain amounts that depend on the signal intensity, $I_s = P_s/A_{eff}$ (where P_s is the coupled optical power).

Moreover, thermal phenomena cannot be neglected. In fact, TPA, FCA and also possible linear surface absorption (at the interface core/cladding Si/SiO$_2$) heat up the waveguide and n_{eff} increases, due to the thermo-optic effect.

Furthermore, given the high $\chi^{(3)}$, another effect may be present, known as the Kerr-effect. In practice, the refractive index n increases with the amount $\Delta n = n_2 I_s$. If the signal intensity is time variant, the incoming optical signal modulates itself. This phenomenon is known as Self-Phase Modulation (SPM).

If the signal is spectrally broad (i.e., composed of multiple harmonics), other effects occur, such as Cross-Phase Modulation (XPM) and Four Wave Mixing (FWM).

When the topology of the considered device includes resonant structures, such as MRRs, the magnitude of those phenomena is scaled by the intensity enhancement factor [2, 5]. This means that the features of the PIC are severely impaired. In particular, the spectral response is distorted, the in-band isolation is reduced and/or bistable behaviour may arise.

5.2.2 Practices to Counteract Nonlinearities

A few strategies have been proposed to counteract nonlinear effects in silicon photonic circuits, based on MRRs.

Suitable polymer coatings have been synthesized and effectively exploited, achieving a power insensitivity of a single MRR transfer function [6, 7].

However, these polymers are typically non-CMOS compatible, increasing the complexity of the fabrication process and depriving the silicon PIC of one of its most appealing features.

Alternatively, one could rely on the careful design of the photonic nanowire, from both physical and geometrical perspectives [8], to play with nonlinear effects and achieve mutual cancellation among them.

This strategy may lead to a more complex fabrication processes and these physical-geometrical constraint may also be in contrast with the other requirements of PIC design and/or fabrication.

Finally, none of the mentioned methods has been applied to a complex structure such as a high-order MRR based filter.

5.3 Dynamic Mitigation of Nonlinear Effects

5.3.1 Evidence of Nonlinear Effects in a Tuneable Optical Add/Drop Multiplexer

The setup exploited to evaluate the nonlinear effects in the DUT is shown in Fig. 5.1.

The transmitted signal is the 200 Gbit/s DP-16QAM (centered at 1539.8 nm), already described in Sect. 2.4 [10]. The transmitter is cascaded with an EDFA and an in-line VOA to finely control the amount of optical power coupled to the PIC, which is, by now, the structure shown in Fig. 2.15a, composed of PSR-straight waveguide (5 mm long)-PRC.

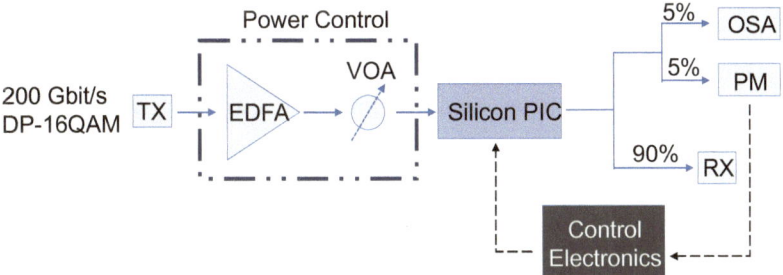

Fig. 5.1 Setup exploited to perform measurements on the PIC with high power signal. *Silicon PIC* might be the test structure in Fig. 2.14 (polarization diversity filter) or that in Fig. 2.15 (PSR-straight waveguide–PRC). Reprinted with permission from [9] Optica Publishing Group

This test structure is useful to evaluate and discriminate all the nonlinear effects that arise in this polarization diversity platform.

At the output of the PIC 90% of the optical power is routed to the coherent receiver, while the other portion of the signal is equally split between a PM and an OSA.

The coupled optical power at the input of the structure is progressively increased. The in-waveguide average power was estimated to be between 8 dBm and 20 dBm. The output spectra and the integral output power have been measured and reported in Fig. 5.2(a) and (b), respectively. There is no spectral distortion, or evidence of power-dependent loss. These results demonstrate (quite surprisingly) the absence of TPA and FCA, in the considered power range. Thus, the number of generated free-carriers should be quite small, and also FCD could be neglected. This could be due to the following reasons:

- the waveguide is not long enough. Considering a reasonable TPA coefficient $\beta_{TPA} = 6(\pm 2) \cdot 10^{-12}$ m/W [3], the extra loss, only due to TPA, is given by $\alpha_{TPA} = \beta_{TPA} P_s / A_{eff} = 3$ Np/m ≈ 27 dB/m. The maximum coupled power is 100 mW, however each waveguide is hit by a halved average power (P_s, since the overall optical power is split, on average, in the two State of Polarization (SOP)s). In a 5-mm-long waveguide the integral loss is 0.13 ($\pm$ 0.05) dB. In the best case it means only 0.08 dB (per waveguide);
- within the fabrication, H_2 annealing (at 800° C) was performed. Reflow usually reduces the lifetime of free-carriers and consequently the nonlinear optical losses [11].

The thermal effects seem to also be negligible.

Thus, we narrow it down, to just the Kerr-effect.

To predict the impact of Kerr-nonlinearities on the single TOADM channel (single filter) response, we evaluated the enhancement factor S, within the passband of each MRR, by means of a numerical (photonic) circuit simulator. S factors are wavelength dependent, and are reported in Fig. 5.2. Since the filter has been implemented through a Vernier scheme, the resonators have different Q-factors, and hence different S factors. For the two innermost rings, S is close to 10 at the central frequency, while it is >24 at the band edges. For outer rings, S is always <8. This means that the nonlinearities are stronger in the two central MRRs, leading to a spectral response distortion (Fig. 5.3).

Therefore, the Kerr-effect induces a phase shift in the resonators given by [12]

$$\phi_{NL,r} = \frac{S^2}{2} \frac{2\pi}{\lambda} n_2 \frac{P_s}{A_{eff}} L_r, \tag{5.1}$$

Fig. 5.2 a PSD of the 200 Gbit/s signal coming out from the structure in Fig. 2.15a, for different values of the coupled (average) input power, shown aside. **b** In-Out power transmission, which does not change with coupled input power. Adapted with permission from [9] ©Optica Publishing Group

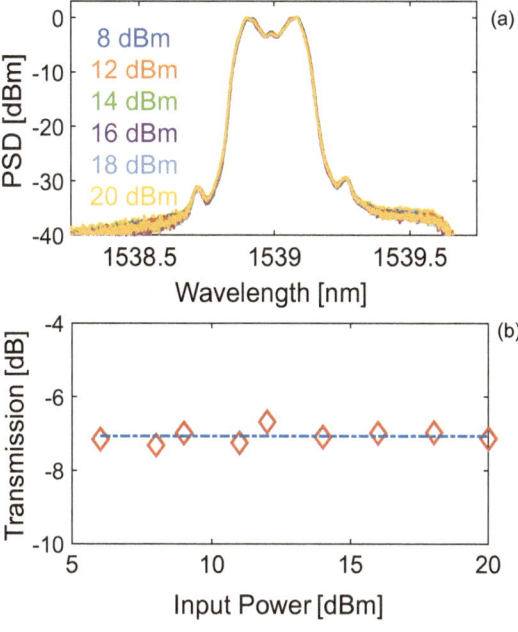

Fig. 5.3 Simulated enhancement factor versus wavelength, for the four MRRs, normalized to the input power. 3dB-Bandwidth (B) is indicated. Reprinted with permission from [9] ©Optica Publishing Group

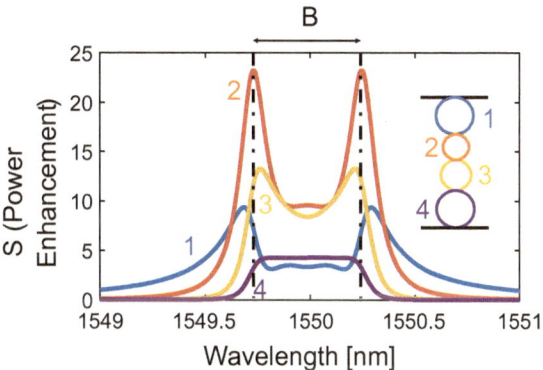

where L_r is the single MRR length (having different radii, it is comprised between 51 and 90 μm) and P_s is the optical power in the input waveguide[1].

This phase shift is responsible for the red-shift (the positive variation of n_{eff}) of the ring resonances, given by the well known equation:

$$\delta f_r = FSR_r \frac{\phi_{NL,r}}{2\pi}, \tag{5.2}$$

---

[1] Unless otherwise specified P_s represents the average optical power in the input bus waveguide, since in this work, a modulated signal has been used as a probe signal. The same for the optical power values reported in the pictures.

where FSR_r is the Free Spectral Range of the single MRR. Therefore, the frequency shift is much stronger in the two inner rings, than in the edge rings. By the way, the central rings also determine the bandwidth of the filter itself.

According to Eqs. (5.1) and (5.2), the induced nonlinear phase shift is approximately 32 mrad for the second and third rings (at the center of the band, approximately 1539.8 nm), which corresponds to approximately 6 GHz frequency shift, with their FSR between 1.2 and 1.4 THz.

This frequency shift has also been confirmed by the measurements. Exploiting the setup described at the beginning of this paragraph and using as DUT the polarization independent sample described in the Sect. 2.3.4 and shown in Fig. 2.14, we measure the spectra of the 200 Gbit/s signal, coming out from the output ports, for increasing coupled power. The chip is stabilized in temperature, at 25 °C, by means of a TEC and it has been previously tuned to reach its nominal transfer function (passband central wavelength matched to central wavelength of the signal, 1539.8 nm), as in Fig. 2.28. The device is not actively controlled.

The signal spectra coming out from Drop port [see Fig. 5.4a] lose symmetry and seem to be progressively sliced on the left side (redshift), up to 5.5 GHz, when the power in the

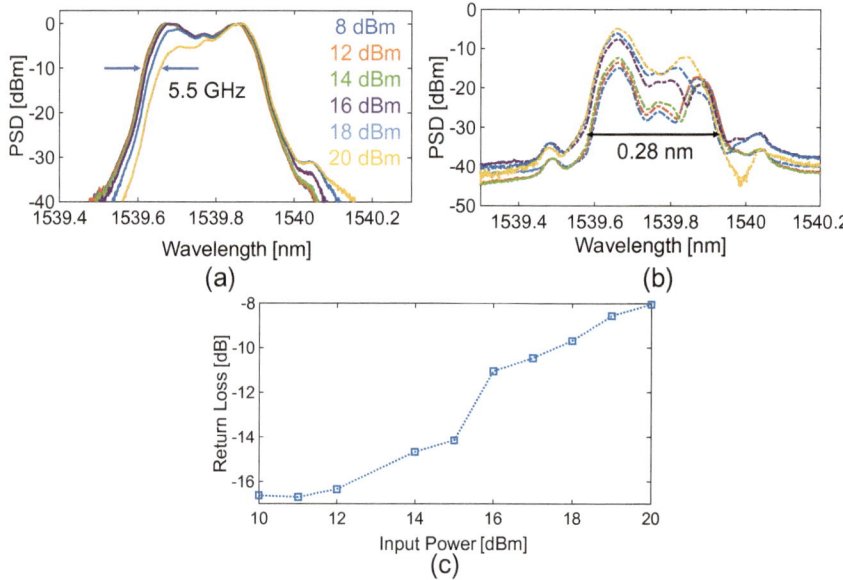

Fig. 5.4 PSD of the 200 Gbit/s signal coming out from the structure (without any active control loop) in Fig. 2.14, for different values of coupled input power. In-Drop (**a**) and In-Through (**b**) conditions are accounted. The bandwidth narrowing is highlighted. **c** Return Loss averaged on 0.28 nm around signal central wavelength, passing from $\approx 17\ dB$ to just $\approx 8\ dB$, as input power increases. Adapted with permission from [9] ©Optica Publishing Group

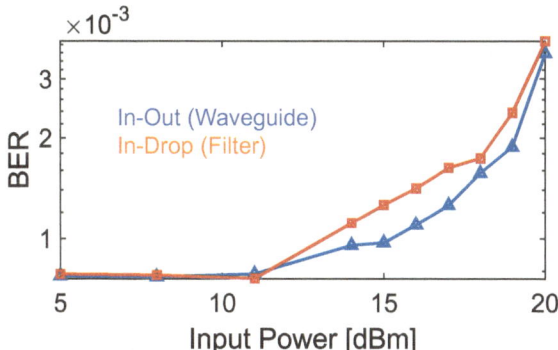

Fig. 5.5 Pre-FEC BER of the 200 Gbit/s signal, against the coupled input power. The blue curve is referred to the In-Out transmission of the structure in Fig. 2.15a, while the orange curve is referred to In-Drop transmission (without an active control loop) of Fig. 2.14. The transmission quality definitely worsens as P_s increases, however up to 20 *dBm* the BER is below the FEC threshold. Adapted with permission from [9] ©Optica Publishing Group

waveguide reaches 20 dBm. The sliced portions of the spectra are redirected to the other output (Through port), causing optical crosstalk.

The spectra collected at the Through port (for the same coupled power range) have also been collected, and reported in Fig. 5.4b. It should be noted that again the spectral symmetry is lost and the return loss progressively decreases, passing from ≈ 17 dB (in the linear regime) to just 8 dB (when the coupled input power, $P_s = 20$ dBm), as shown in Fig. 5.4c. The *RL* has been computed considering a spectral width of 0.28 nm, around the central wavelength of the signal.

For both outputs, the nonlinear effects become relevant for a coupled power >14 dBm.

The evaluation of the impact of these nonlinear phenomena on the transmission quality is also a matter of interest. We evaluate the behaviour of the (Pre-FEC) BER with respect to P_s, for the two polarization diversity structures: straight waveguide and complete filter (In-Drop condition), having the same total geometrical length, approximately 5 mm. The OSNR is kept constant and always >30 dB. The results are reported in Fig. 5.5. The blue curve refers to the straight waveguide, while the orange curve refers to the filter. The BER gradually increases with P_s because the signal experiences a Kerr-induced SPM, affecting the received constellation. Furthermore, the nonlinear spectral distortion gives a BER-penalty higher than 3 dB (for $P_s > 12$ dBm), which severely impairs the transmission quality.

5.3.2 Nonlinearities Mitigation in Tuneable Optical Add/Drop Multiplexer

The nonlinear effects have detrimental consequences on the DUT, at both the device and system levels. In this section we propose and discuss an approach to dynamically mitigate them, by using the automatic feedback loop (i.e., thermally controlling each MRR) already presented in Chap. 4.

The technique has a number of benefits, such as:

- local action (i.e., only the elements suffering from nonlinear distortion are selectively tuned). This is particularly important for complex and high-density PICs, since high power hits only a single portion of the circuit (for example a single channel of a TOADM);
- complete blindness. The procedure works with no a priori information about the topology of the PIC, the platform it is made on, the involved degrees of freedom (phase shifters, tuneable couplers [13], etc.) or the kind of nonlinear phenomena it experiences (nonlinear phase or amplitude variations, provided that the PIC is equipped with the right type of actuators);
- scalability to more complex circuit topology and to other circuit technologies (such as III–V compounds);
- working with standard waveguides/circuits (i.e., waveguide geometry should not be changed and there are no extra compounds surrounding it).

The proposed method exploits again a gradient descent algorithm that minimizes the power coming from the Through port of the photonic circuit and sets (accordingly) the driving signals for the thermo-optic actuators. The characteristics of the exploited electronics and tuning scheme can be found in Chaps. 3 and 4, respectively. Similar to what has been discussed in Chap. 4, these signals are weighted to proper coefficients to cancel thermal crosstalk and are adaptively controlled to guarantee the stability of the control loop and speed up its convergence.

The block scheme of the polarization insensitive filter equipped with its own control electronics is reported in Fig. 5.6.

The active control loop is turned on and the power of the coupled signal is progressively increased. At the same time, spectral measurements are performed and reported in Fig. 5.7a, b, for Drop and Through ports, respectively.

The RL, always evaluated considering a spectral width 0.28 nm [see Fig. 5.7c, where it is shown together with (orange curve) and without (blue curve) any active control], appears to be almost power-independent. The spectral symmetry is recovered in both conditions, and in particular, the signals at the Drop port are not sliced, nor is the bandwidth narrowed.

Concerning the BER performance, In-Drop transmission has been evaluated, with the active control loop on. The (Pre-FEC) BER versus P_s curve is reported in green in Fig. 5.8. For convenience, in this picture also BER versus P_s curves in the case of a straight waveguide

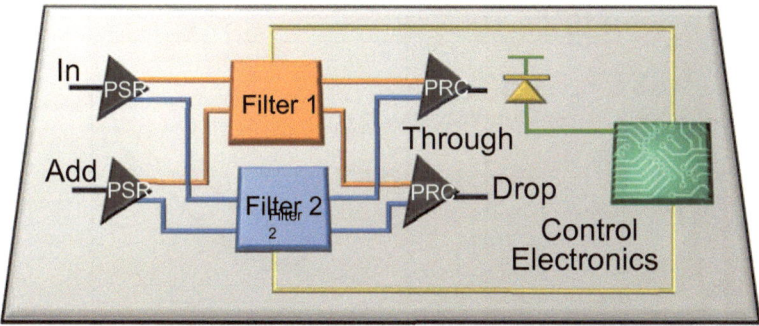

Fig. 5.6 Block scheme of the polarization diversity filter, with control electronics. Reprinted with permission from [9] ©Optica Publishing Group

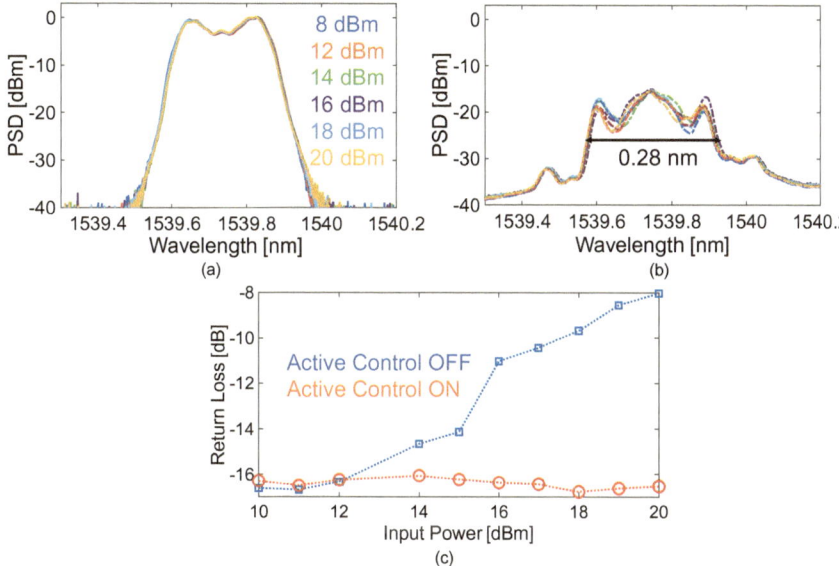

Fig. 5.7 PSD of the 200 Gbit/s signal coming out from the structure (with active control loop turned on) in Fig. 2.14, for different values of coupled input power. In-Drop (**a**) and In-Through (**b**) conditions are accounted for. **c** Return Loss averaged on the value of 0.28 *nm* around the signal central wavelength, is almost constant for all the accounted values of input power, approximately 17 *dB*. Adapted with permission from [9] ©Optica Publishing Group

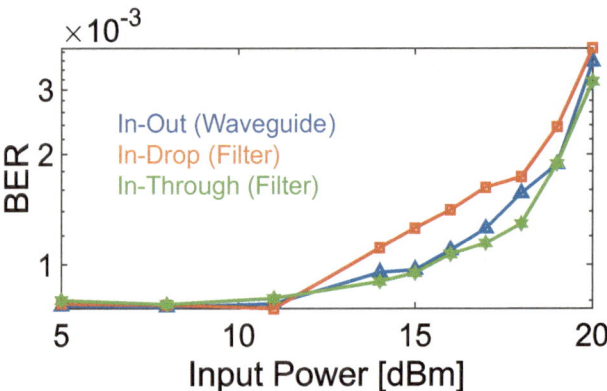

Fig. 5.8 Pre-FEC BER of the 200 Gbit/s signal, against the coupled input power. The blue curve refers to the In-Out transmission of the structure in Fig. 2.15a, and the orange and green curves refer to In-Drop transmission (without and with active control loop on, respectively) of Fig. 2.14. The transmission quality definitely worsens as P_s increases, however up to 20 dBm BER is below the FEC threshold. Reprinted with permission from [9] ©Optica Publishing Group

(blue) and in the case of control loop turned off (orange) are also reported. The results show that the power-dependent penalty is removed by the active control (green and blue curves overlap), while SPM is always present and depends on the silicon propagation phenomenology [14], not on the DUT topology.

In conclusion, since the strategy for the compensation relies on phase actuators, the nonlinear effects we are counteracting are (mainly) nonlinear phase shifts.

5.4 LookUp Table Update

To achieve the results discussed in the previous section, the working points of thermo-optic actuators have to be slightly modified.

Clearly, the two central rings require a more significant correction, approximately -1 mW in the electrical power domain, compared to that imposed on the external rings, which is more than one order of magnitude lower (≈ -0.1 mW).

In Fig. 5.9 electrical power corrections are plotted against P_s, for all the MRR (TCs' actuators do not require any extra correction).

A negative sign in front of these corrections is required. The electrical power absorbed by the loads must be reduced (negative variation of n_{eff}), in order to compensate for the Kerr-induced redshift.

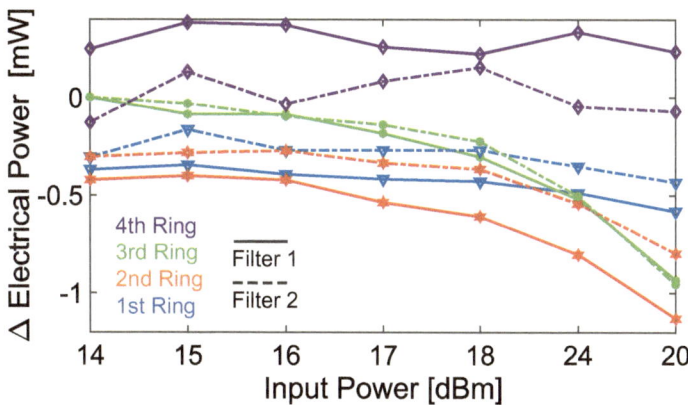

Fig. 5.9 Electrical power correction assigned to each MRR with respect to the coupled input power. The values are correctly <0 mW, to counteract red-shift

Passing from the linear to the nonlinear regime requires the execution of approximately 30 iterations of the control loop, before reaching convergence (i.e., tens of milliseconds, according to the considerations made before, about the thermal cross-talk dynamics and the thermo-optic effect time constant).

The experiment was repeated several times, even introducing random perturbations (±5% standard deviation) on the (linear regime) driving signals of the MRR actuators, before executing control. In all the considered cases, the same set of driving signals was obtained, at convergence.

Hence, the presence of high-power signals, which may trigger nonlinear phenomena, is a situation that requires LUT update, as much as temperature variations, modification of incoming signal PSD or new users' needs.

The updated LUT is reported in Table 5.1, for the wavelength channel of interest (1539.8 nm).

The spectral response of the filter in the linear domain, applying these LUT entries, is reported in Fig. 5.10a, b (for Through and Drop ports, respectively). It is quite far away from the nominal (desired) transfer function. Therefore, the DUT could be predistorted and assume this shape, ready to accept high-power signals and accommodate them.

Table 5.1 Updated LUT for the single channel-double polarization filter under test, for different coupled optical power values

Building block	@8 dBm [mW]	@10 dBm [mW]	@12 dBm [mW]	@16 dBm [mW]	@18 dBm [mW]	@20 dBm [mW]
Top MZI–Filter 1	4.16	4.16	4.16	4.18	4.20	4.22
Ring 1–Filter 1	11.98	11.68	11.66	11.57	11.53	11.48
Ring 2–Filter 1	7.96	7.56	7.42	7.28	7.14	7.00
Ring 3–Filter 1	4.58	4.18	3.64	3.90	3.76	3.62
Ring 4–Filter 1	8.08	7.78	7.79	7.68	7.63	7.58
Bottom MZI–Filter 1	11.08	11.08	11.08	11.10	11.14	11.16
Top MZI–Filter 2	4.26	4.26	4.26	4.28	4.30	4.32
Ring 1–Filter 2	13.03	12.73	12.68	12.63	12.58	12.53
Ring 2–Filter 2	6.06	5.66	5.52	5.38	5.24	5.10
Ring 3–Filter 2	4.01	3.61	3.47	3.33	3.19	3.05
Ring 4–Filter 2	9.81	9.51	9.46	9.41	9.36	9.31
Bottom MZI–Filter 2	15.06	15.06	15.06	15.10	15.12	15.14

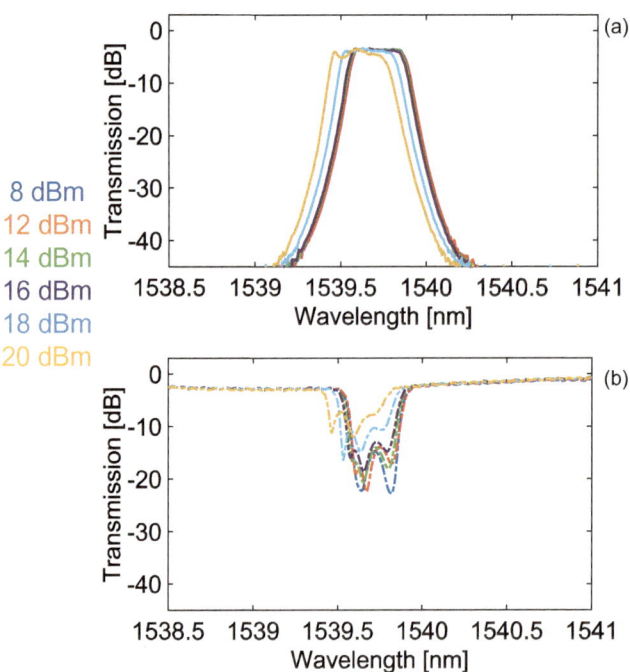

Fig. 5.10 Spectral responses (in the linear regime) imposing the computed LUT on the polarization diversity filter. Correctly, these functions are slightly shifted toward shorter wavelengths

References

1. I. Demirtzioglou, C. Lacava, D. J. T. Kyle R. H. Bottrill, G. T. Reed, D. J. Richardson, and P. Petropoulos, "Frequency comb generation in a silicon ring resonator modulator," *Optics Express*, vol. 26, no. 2, 2018.
2. A. C. Turner, M. A. Foster, A. L. Gaeta, and M. Lipson, "Ultra-low power parametric frequency conversion in a silicon microring resonator," *Optics Express*, vol. 16, no. 7, 2008.
3. J. Leuthold, C. Koos, and W. Freude, "Nonlinear silicon photonics," *Nature Photonics*, vol. 4, no. 8, 2010.
4. Q. Lin, O. J. Painter, and G. P. Agrawal, "Nonlinear optical phenomena in silicon waveguides: modeling and applications," *Optics Express*, vol. 15, no. 25, 2007.
5. M. Novarese, S. R. Garcia, S. Cucco, D. Adams, J. Bovington, and M. Gioannini, "Study of nonlinear effects and self-heating in a silicon microring resonator including a Shockley-Read-Hall model for carrier recombination," *Optics Express*, vol. 30, no. 9, 2022.
6. L. He, Y.-F. Xiao, C. Dong, J. Zhu, V. Gaddam, and L. Yang, "Compensation of thermal refraction effect in high-Q toroidal microresonator by polydimethylsiloxane coating," *Applied Physics Letters*, vol. 93, no. 20, 2008.
7. S. Grillanda, V. Raghunathan, V. Singh, F. Morichetti, J. Michel, L. Kimerling, A. Melloni, and A. Agarwal, "Post-fabrication trimming of athermal silicon waveguides," *Optics Letters*, vol. 38, no. 24, 2013.
8. L.-W. Luo, G. S. Wiederhecker, K. Preston, and M. Lipson, "Power insensitive silicon microring resonators," *Optics Letters*, vol. 37, no. 4, 2012.
9. M. Petrini, M. Milanizadeh, F. Morichetti, and A. Melloni, "Dynamic mitigation of nonlinear effects in a silicon photonic add-drop filter," *Optics Letters*, vol. 46, no. 19, 2021.
10. "Jabil Photonics, CFP2-DCO, https://www.jabil.com/industries/photonics/photonics-resources.html."
11. Y. Liu and H. K. Tsang, "Raman gain in helium ion implanted silicon waveguides," in *Conference on Lasers and Electro-Optics (CLEO) 2006*, IEEE, 2006.
12. A. Melloni, F. Morichetti, and M. Martinelli, "Linear and nonlinear pulse propagation in coupled resonator slow-wave optical structures," *Optical and Quantum Electronics*, vol. 35, no. 4/5, 2003.
13. P. Orlandi, F. Morichetti, M. J. Strain, M. Sorel, A. Melloni, and P. Bassi, "Tunable silicon photonics directional coupler driven by a transverse temperature gradient," *Optics Letters*, vol. 38, no. 6, 2013.
14. X. Liu, J. B. Driscoll, J. I. Dadap, R. M. Osgood, S. Assefa, Y. A. Vlasov, and W. M. J. Green, "Self-phase modulation and nonlinear loss in silicon nanophotonic wires near the mid-infrared two-photon absorption edge," *Optics Express*, vol. 19, no. 8, 2011.

Testing of Delay Lines Breaking Bandwidth-Delay Constraint

6

6.1 Introduction

In this Chapter another class of devices is considered, Delay Lines. So far, we have discussed only Infinite Impulse Response (IIR) based devices (e.g., PIC implemented by means of several MRRs).

In these pages, instead, we present two different delay line topologies, implemented through the use of MZIs, which are Finite Impulse Response (FIR) architectures. In so doing, we generalize as much as possible the testing treatment, dealing with different building blocks.

Moreover, like the Add/Drop filter, these other PICs have some interesting peculiarities, that deserve to be deepened. In particular, the architectures we propose are able to overcome the traditional Delay-Bandwidth Constraint.

Similar to the TOADM operating in a WDM environment, these delay structures are made in Silicon Photonic platform and are programmable, by using thermo-optic phase shifters.

Thus, the electronic platform to perform electrical testing and optical tuning is the same as that exploited for photonic filters (see Chap. 3).

Furthermore, some of the techniques described in Chap. 4, may become also important in this scenario and, by merging them with new strategies, specifically developed for these new DUTs, some testing recipes for delay lines could be effectively synthesized.

The Chapter is organized as follows:

- In Sect. 6.2 the applications and limits of optical delay line circuits are discussed.
- In Sect. 6.3 the first topology (composed of 4 nested MZIs) overcoming these limits is described; its operation is validated, its testing procedure is synthesized and further developments are proposed.

© The Author(s), under exclusive license to Springer Nature Switzerland AG 2025
M. Petrini, *Mixed-Signal Generic Testing in Photonic Integration*, Synthesis Lectures on Digital Circuits & Systems, https://doi.org/10.1007/978-3-031-60811-7_6

- In Sect. 6.4, following the path of Sect. 6.3, a second topology (composed of 2 cascaded MZIs) is described, its operation is validated and its testing procedure is synthesized.

6.2   Application and Limits of Optical Delay Lines

Optical true time delay lines are a key element for a number of different applications. Among them, we can certainly find optical communications, for example for data synchronization [1], in Optical Time Domain Multiplexing (OTDM) systems [2], and for data buffering [3] in Optical Computing systems.

Recently delay lines have been proven to be essential for microwave photonics systems [4] and optical beam forming architectures for phased-array antennas [5] [6]. In particular, if properly used, they enable for an increase of network bandwidth and prevent beam squinting.

True time delay devices can be implemented in different technologies such as fiber optics [7] or free-space [8]. However, photonic integration offers some attractive features for this purpose, such as a small footprint, low power consumption and reduced cost.

Many different delay line topologies, implemented by means of integrated photonic filters [9], can be found in the literature. In general, all of them could be clustered into two categories, i.e., implemented by means of resonant or non-resonant building blocks.

In the first set, we can find, for example, photonic crystals and ring resonators [10]. In particular, resonance-based architectures enable group delays up to hundreds of ps, in a relatively compact footprint, but at the same time they show narrow bandwidth and quite large chromatic dispersion [11]. The bandwidth can be enlarged by combining several MRRs in Coupled-Resonators Optical Waveguide (CROW) structures. The control and calibration procedure for these architectures made of a large number of cavities may be quite complex, and, as we discussed in Chap. 5, nonlinearities might not be neglected [12].

In the latter, instead, we can find structures having a larger operational bandwidth, (typically) a simpler and more accurate tuning strategy, more resilience to nonlinearities and a group delay less dependent on wavelength. These appealing features are counterbalanced by a larger footprint and a shorter group delay (since the delay is proportional to the optical path of the traveling signal).

Concerning this category, in the literature there are examples of delay lines implemented via high-extinction-ratio switches that route the optical wave through different paths, with different lengths. However, in so doing, the total group delay can assume only a discrete set of values [13].

The continuous tuneability of the introduced delay is a quite appealing feature, and it has been achieved already, building a MZI-based delay line (sketched in Fig. 6.1, where Input-Output and TCs are highlighted), equipped with TCs [14].

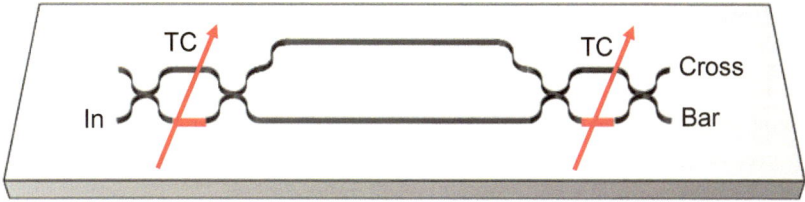

Fig. 6.1 Schematic of a MZI, implemented by means of two TCs, that realizes a simple continuously tuneable delay line

The maximum time imbalance, between the shortest and the longest arm of the interferometer (at the central frequency) is given by

$$T_{MZ} = \frac{n_g \Delta L}{c} = FSR^{-1},\tag{6.1}$$

where ΔL is the geometrical imbalance of the MZI itself.

Hence the maximum group delay, that the optical wave experiences, is

$$\tau_{max} = \tau_0 + T_{MZ},\tag{6.2}$$

where τ_0 is the delay introduced by the shortest arm of the MZI.

In the mentioned reference, [14], it is also proven that the signal group delay (τ) is a linear function of the coupling coefficients of the TCs (always assumed to be exactly at the same working point, i.e., $K = K_1 = K_2$) with which the delay line is equipped.

Thus, τ can be written as

$$\tau = \tau_0 + (K)(T_{MZ}).\tag{6.3}$$

According to Eq. (6.3), the fine control of K (continuously tuneable between 0 and 1) allows for continuous and precise control of τ.

However, all the presented structures, regardless of their nature (IIR or FIR) are subjected to the constraints on propagation insertion losses and the delay-bandwidth product.

First, the propagation losses are given by $IL = \alpha \tau c / n_g$. There is no dependence on circuit topology. Hence, IL/τ ultimately depends only on the quality of the waveguides (in terms of α/n_g ratio).

Furthermore, a longer delay could be achieved at the expense of narrower bandwidth and conversely, a shorter delay increases the bandwidth. In short, $\tau B_{3\,dB}$ is limited.

For example, in a (single) MZI-based delay line, with 50:50 directional couplers, $\tau B_{3\,dB}$ is exactly equal to $\frac{1}{4}$; while this figure of merit is three times lower in an all-pass MRR.

In the following sections, we propose two different topologies to overcome this limitation. In addition, by means of proper time-domain and spectral measurements, we characterize them. In conclusion, using some of the results previously achieved in this book, we synthesize a technique to effectively test them.

6.3   Tuneable Coupler Characterization

The considerations made in the previous paragraph suggest the importance of TCs in the implementation of continuously tuneable optical delay lines. In our designs this building block is crucial and widely used. For this reason, its description and characterization are important, before proceeding with further considerations.

From a circuit perspective, the TCs are implemented by means of balanced MZI (i.e., whose geometrical imbalance is 0 mm), with nominally identical 50:50 directional couplers. Such a device is sketched in Fig. 6.2a.

By means of a phase shifter (placed above one of the two arms), the power splitting ratio between the two outputs could be finely controlled, from 0 (bar condition) to 1 (cross condition).

Tuneable couplers (as all the waveguides and sub-structures of the proposed PICs), are made in a 220nm-commercial silicon photonic platform [16]. The actual realized device is shown in Fig. 6.2b. The adopted waveguide is channel shaped (500 nm by 220 nm, with a bottom oxide 3 μm thick) and its cross-section is in the inset Fig. 6.2a.

Resistive thermo-optic phase shifters are effectively exploited for the actuation (i.e., fine control of the coupling ratio). These heaters are 85μm long and are placed 700 nm above the waveguide core. Their nominal resistance is approximately 340 Ω.

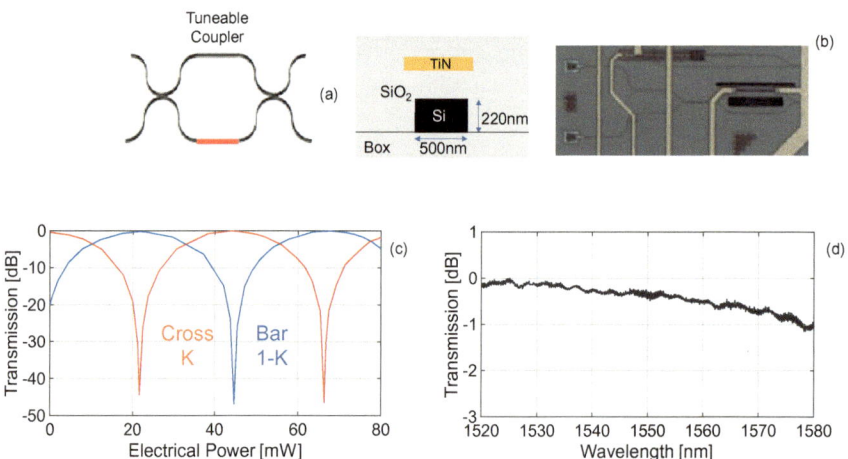

Fig. 6.2 a Schematic of the considered TC (with waveguide cross-section), implemented by means of a balanced MZI (with 50:50 directional couplers). It can be controlled using the heater above one of the arms. **b** Top-view microphotograph of the implemented TC. **c** Coupling ratio against the electrical power absorbed by the actuator. The power to be dissipated to achieve a 2π phase shift is $\approx 40\ mW$. **d** In-Out spectrum of the TC, when tuned to reach full cross condition (i.e., $K = 1$) around 1520 nm. Its $1dB$-bandwidth is as large as 60 nm. Adapted with permission from [15] ©optica publishing group

This building block, such as the more complex structure we describe in the next sections, is optically accessed by means of Input-Output grating couplers.

A narrowband signal (monochromatic light source, at 1550 nm) is injected in the input port of the device and a photodetector is coupled to the bar port and cross port, one at a time. An analog voltage ramp (between 0 V and 5 V, in 1 s) is provided to the heater (by means of a SMU) and, while reading the absorbed current, the output of the photodetectors was sampled (both quantities with a sample rate of 1 MHz). These data are summarized by the plot in Fig. 6.2c, which shows the behaviour of the optical coupling ratio against the electrical power provided to the actuator (for both bar and cross ports).

When the actuator is not driven, the TC is in quasi-cross condition (as expected by design) and a 2π shift occurs in approximately 40 mW.

The couplers implemented by means of balanced MZIs show less wavelength dependence, than standard directional couplers.[1]

However, when the directional couplers of the balanced-MZI are very far from 3 dB condition, the so-implemented TC might not reach complete cross condition.

The behaviour of the fabricated building blocks has been investigated by proper spectral measurements. A monochromatic light source (approximately 1520 nm) is coupled to the input port of the TC, and the cross condition is achieved, by controlling the phase shifter. Then, by means of a TLS synchronized with an OSA, the spectral response (In-Cross) of the MZI is measured. The outcome is shown in Fig. 6.2d. The cross state not only can be achieved at approximately 1520 nm (30 nm from 1550 nm, the wavelength for which directional couplers are designed), but also 1 dB bandwidth is as large as 60 nm, which is sufficient for the desired application. The cross condition could also be achieved around 1580 nm by re-tuning the TC. Notably, the grating couplers frequency response has been de-embedded from the measured curves.

6.4   Four-Stage Delay Line

6.4.1   Topology

The first proposed architecture is in the schematics in Fig. 6.3a. It is composed of four nested imbalanced MZIs, configured in such a way that the longer arm of the N-th stage is the reference arm of the (N+1)-th MZI. Every single imbalanced interferometer is implemented by using two TCs, whose features are those described in the previous paragraph. The TCs in the picture (and from now on) are labeled as $K_{A,i}$ and $K_{B,i}$, where A and B identify the

---

[1] It can be proven through trivial numerical simulations that a balanced MZI could reach bar and cross condition even if implemented by means of a pair of directional couplers, whose coupling ratio is intentionally perturbed by $\pm 10\%$ with respect to their nominal value. As discussed in the main text, a directional coupler may not match design specifications due to fabrication imperfections or its own wavelength-dependence.

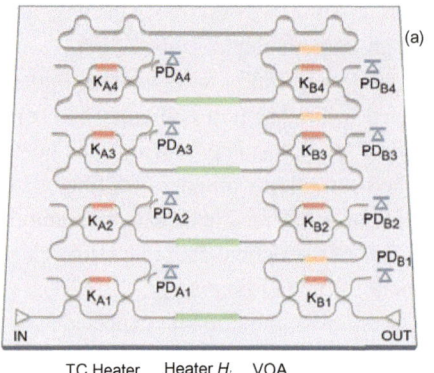 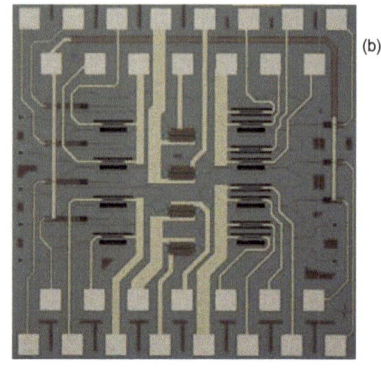

Fig. 6.3 **a** Scheme of the first proposed delay line. $K_{x,i}$ and $PD_{x,i}$ (with $x = A, B$ and $i = 1, 2, 3, 4$) represent TCs and PDs. The heaters controlling TCs are highlighted in red, and those above the imbalances are in orange. The VOAs are highlighted in green. The In-Out Grating Couplers are also shown. **b** Top-view microphotograph of the implemented PIC. Reprinted with permission from [15] ©optica publishing group

input or output column, while $i = 1, 2, 3, 4$. In the following we will assume $K_{A,i} = K_{B,i}$ (i.e., TCs belonging to the same interferometer tuned to the same working point).

The coupling ratio of TCs (programmable by means of heaters, highlighted in red in the picture) can be precisely estimated by means of integrated Ge-PDs[2] [16] placed at one of the TCs' outputs, directly or through 10% tap (these, coherent with TCs are labeled as $PD_{A,i}$ and $PD_{B,i}$).

The MZI imbalances, (geometrical) $\Delta L = 1.7$ mm, are equipped with other heaters (highlighted in orange, 85 μm long) to shift the central frequency of the PIC and match the carrier wavelength of the input signal.

The VOAs,[3] whose importance will be clear in the following, are integrated in the MZI central waveguides (highlighted in green, in the picture) and complete the whole architecture, shown in the top-view microphotograph, Fig. 6.3b. The optical access is performed by means of grating couplers. The footprint of the realized PIC is 1.2 mm by 0.6 mm, excluding electrical pads.

---

[2] The same PDs discussed in Chaps. 2 and 3.

[3] In the vicinity of VOAs, the waveguide becomes rib-shaped and tapers allow a smooth transitions between the two different cross-sections. VOAs are implemented by means of p-i-n junctions. The p-doped and n-doped regions (both 10^{20} cm^{-3}) are placed 900 nm from waveguide core.

6.4.2 Device Operation

According to Eq. (6.1), each stage can provide a maximum extra-group delay (with respect to that of the reference path, τ_0) of $T_{MZ} = 22.5$ ps, with $n_g = 3.89$. The FSR is 44.4 GHz, while the minimum $B_{3\,dB} = FSR/2 = 22.2$ GHz (this condition is reached when both TCs of the same stage are at the 3dB-split condition). Since (up to) 4 stages are available, the maximum group delay is $\tau_{max} = \tau_0 + T_{D,max} = \tau_0 + 4T_{MZ} = \tau_0 + 90$ ps, with $T_{D,max}$ being the maximum extra group delay of the whole delay line (with respect to that of the reference path).

The actual working principle is now described.

If the desired extra group delay (T_D) is between 0 ps and T_{MZ}, $K_{A,2}$ and $K_{B,2}$ should be set to 0 (bar condition, i.e., $K_{A,2} = K_{B,2} = 0$), while $K_{A,1}$ and $K_{B,1}$ should be set to the value T_D/T_{MZ} to obtain the desired delay. In this situation only the first stage is actually operative and the upper stages are completely isolated.

If the desired extra group delay (T_D) is between T_{MZ} and $2T_{MZ}$, $K_{A,3}$ and $K_{B,3}$ should be set to 0 to isolate the third and fourth stages, and $K_{A,1}$ and $K_{B,1}$ should be set to 1 (both in cross condition). At the same time $K_{A,2}$ and $K_{B,2}$ should be set to the desired working point, i.e., $(T_D - T_{MZ})/T_{MZ} = T_D/T_{MZ} - 1$. In this case only the second stage operates as an actual interferometer, while the first stage contributes with a fixed-length delay, equal to T_{MZ}, induced by its longer arm.

Generalizing the approach to an arbitrary number of stages, it is convenient to introduce the following quantities[4]:

$$M = 1 + \lfloor \frac{T_D}{T_{MZ}} \rfloor, \tag{6.4}$$

$$m = \frac{T_D - (M-1)T_{MZ}}{T_{MZ}} = mant(\frac{T_D}{T_{MZ}}). \tag{6.5}$$

The TCs $K_{A,i} = K_{B,i}$ are set in cross condition, with i between 1 and $M - 1$ (this does not hold when the considered interferometer is the first stage). In so doing they induce a delay equal to $(M - 1)T_{MZ}$. Conversely, $K_{A,M+1} = K_{B,M+1}$, if present, are set to 0, to isolate the upper stages (if it is the case). In this way, the M-th MZI is the only stage that works as an interferometer, and sets the residual delay mT_{MZ}.

The fact that, regardless of the total delay, only one MZI is operative is the key element that permits effectively breaking the constraint on the bandwidth-delay product.

---

[4] *mant* stands for *mantissa*, i.e. fractional part of a number.

6.4.3 Testing Procedure

First, preliminary electrical characterization is performed.[5]

The optical Power-Current-Voltage characteristics of all the sub-elements (every single TC and its own output PD) are serially acquired ($K_{A,i}$ features are measured when $K_{A,i-1}$ is put in cross condition and column B is measured after column A, following reverse order). The I-V curves for all the actuators involved in the fourth stage (two heaters controlling TCs, one phase-shifting heater, and one p-i-n junction) are shown in Fig. 6.4a. For the two TCs, the coupling-voltage characteristics are in Fig. 6.4b (with this method, we measure the bar condition of TCs in column B, i.e., $1 - K_{B,i}$).

This testing stage, if passed, ensures the correct functionality of the DUT.

Following the steps executed for the previously tested PICs, the thermal cross-talk matrix should be estimated. Notably, in this circuit, thermo-optic phase shifters are not closer than $40\,\mu\text{m}$. Moreover, each heater is surrounded by deep trenches, aiding the thermal dissipation. The considerations on $\mathbf{T}$, similar to those carried out in Sect. 3.6, lead to the fact that there is thermal coupling only between H_i and $K_{B,i}$ (with $i = 1, 2, 3, 4$). The actual amount of coupling (for the four pairs) is between 4.7% and 5.5%.

Fig. 6.4 a Current-voltage characteristics of the heaters and p-i-n junction embedded in the PIC (only for the last stage). **b** Voltage-optical Power characteristics of the TCs, constituting the described delay line (only for the last stage)

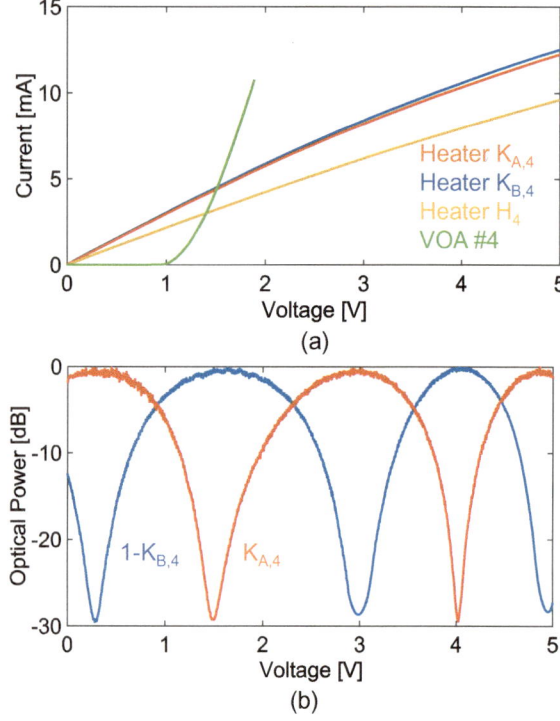

---

[5] Device temperature stabilized at $25\,^{\circ}C$ by means of a TEC.

Once electrical testing is performed and the thermal crosstalk is estimated, a proper LUT can be built, for all the desired values of τ.

The controller described in Chap. 3 has been properly programmed to tune this kind of PIC.

The integrated PDs provide feedback signals to the controller, which executes these steps (which are completely independent from the number of stages):

1. The optical signal is coupled to the PIC and, according to the desired τ, M and m are calculated.
2. $K_{A,i}$ and $K_{B,i}$, for $i = 1, \ldots, M - 1$ are put in cross condition by maximizing (in case column A) or minimizing (in case of column B) the power at their output PD.
3. $K_{A,M}$ is put in bar condition (minimizing its out-PD) and after that, set point $K_{B,M} = P_{B,M}/P_{A,M-1}$ is matched with a simple Proportional Integral (PI) control. Relying on the power ratios, rather than absolute values, is beneficial, since the system eliminates the possible fluctuations of the input optical power.
4. Analogously $K_{A,M} = P_{A,M}/P_{A,M-1}$ is achieved.
5. If this is the case, the upper stages of the delay line must be isolated. To do so, $K_{A,M+1}$ and $K_{B,M+1}$ must be put in bar condition (minimizing their out-PDs).
6. Finally, the power at the PIC output is maximized by using the heater H_M, to correctly center its transfer function (if this is the case, i.e., if $K \neq 0, 1$). However this introduces crosstalk on heater controlling $K_{B,M}$. Thus, few iterations around optima must be performed to match both constraints (the power maximization at the output port and $K_{B,M} = P_{B,M}/P_{A,M-1}$).

The tuning accuracy ultimately depends on the resolution of the actuators control and on the PDs' (and the signal acquisition chain's[6]) reliability.

Exploiting hardware described in previous Chapters, the desired τ with an uncertainty <0.1ps is achieved. This could even be improved, by employing DACs with a higher number of bits (and/or reducing the bandwidth of the TIA).

The duration of the entire testing procedure (electrical characterization and calibration algorithm) is <1.5 s (approximately 600 ms for electrical characteristics, 600 ms for the **T** estimation and 100 ms for the calibration scheme).

In this way, again without performing any spectral measurement, we can assess if the device is suitable from the electrical perspective and if it can reach a specific working point, from the optical point of view.

To prove the effectiveness of this "tuning for testing", we perform a posteriori spectral measurements, by using a TLS synchronized with an OSA. We consider the conditions $M = 1, 2, 3, 4$, with $m = 0, 0.25, 0.50, 0.75, 1$ (the longest path is with $M = 4$ and $m = 1$). The spectral responses are reported in Fig. 6.5a, with different colors, for different m,

---

[6] Composed of TIAs, also acting as active analog filters and ADCs, properly calibrated to ensure the correct power estimation.

Fig. 6.5 Measured amplitude (**a**) and Group delay (**b**) Spectra for the considered delay line with $m = 0, 0.25, 0.5, 0.75, 1$ ($M = 4$ for group delay and $M = 1, 2, 3, 4$ for amplitude). The FSR is 356 pm (44.4 GHz), and the minimum bandwidth is 178 pm (22.2 GHz), for $m = 0.5$. The group Delay behaviour is that of a single MZI, but with an offset of 67.5 ps. Reprinted with permission from [15] ©optica publishing group

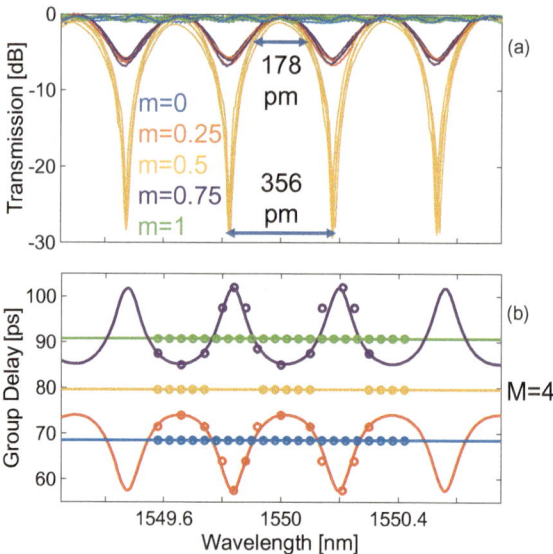

according to the legend. Regardless the operative stage (i.e., chosen M), there is a perfect overlap of transfer functions (with the same m), suggesting that only one interferometer is working at a time. As expected, the FSR is 356 pm, while the minimum $B_{3\,dB} = 178$ pm (for $m = 0.5$).

These frequency responses (with $M = 4$) could be recovered also in spectral portions, far away from 1550 nm, for example around 1520 nm (Fig. 6.6a) and 1580 nm (Fig. 6.6b). According to these results, the PIC shows quite wide operational wavelength range (at least ≈ 60 nm), even larger than the C-band. For $m = 0.5$, the FSR passes from 345 pm (at 1520 nm) to 370 pm (at 1580 nm), i.e., a variation versus frequency of approximately 7%. Moreover, the condition $m = 1$ could always be achieved at the considered wavelengths.

However, to fully ensure the functionality of a delay line (and thus validate the testing technique) measurements on group delay (and in particular its dependence on wavelength, i.e., $\tau_g(\lambda)$) are essential. To do so, the setup in Fig. 6.7 is exploited. The optical carrier is modulated with a 2.5 Gbit/s On-Off Keying (OOK) Non Return to Zero (NRZ) signal. The probe signal, in this case, has a bandwidth 10 times narrower than the minimum bandwidth of the DUT. By using an optical oscilloscope (placed after a fixed gain EDFA and 0.2nm-BPF, for noise filtering), the true time delay (with respect that of the reference path, τ_0) is assessed. The measurements were repeated changing the TLS carrier wavelength, over more than FSR, with a wavelength step $\Delta\lambda = 40$ pm (≈ 5 GHz, more than two times the bandwidth of the probing signal). Experiments were conducted for different values of $m = 0, 0.25, 0.5, 0.75, 1$, keeping $M = 4$.

The results, shown in Fig. 6.5b, show that the behaviour of $\tau_g(\lambda)$ is similar to that expected for a single MZI, but with an offset of ≈ 67.5 ps (i.e., $(M - 1)T_{MZ} = 3T_{MZ}$). Close to the deep notches (with $m = 0.5$) the group delay cannot be defined.

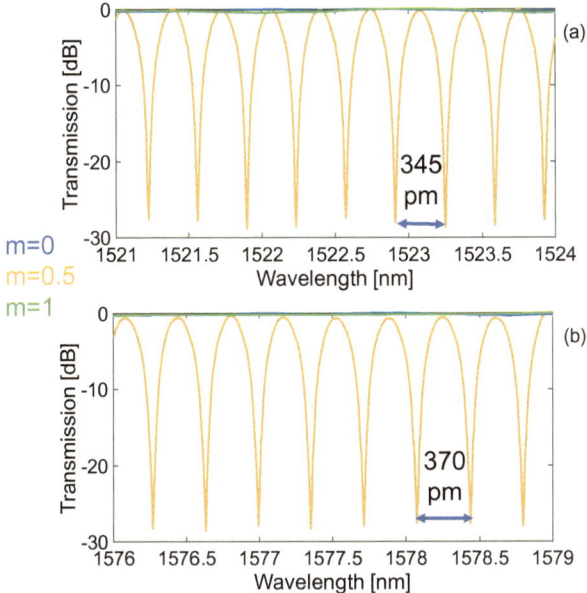

Fig. 6.6 Measured Amplitude Spectra for the considered delay line with $M = 4$ and $m = 0, 0.5, 1$, around 1520 nm (**a**) and around 1580 nm (**b**). The FSR passes from 345 pm to 370 pm. Reprinted with permission from [15] ©optica publishing group

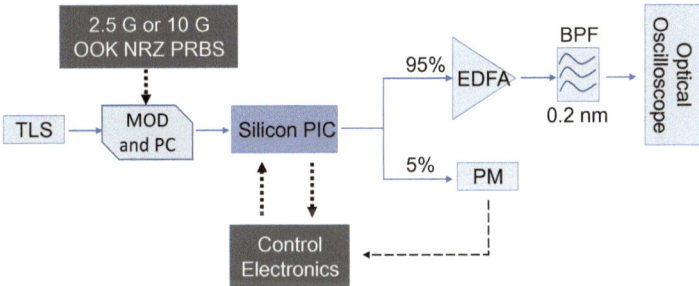

Fig. 6.7 Setup exploited to perform measurement on Group Delay Spectra. The modulator is driven by a 2.5 Gbit/s or by a 10 Gbit/s signal. Reprinted with permission from [15] ©optica publishing group

The estimation of the group delay has also been carried out by using a higher bit-rate signal, 10 Gbit/s OOK NRZ, whose carrier wavelength is matched (and fixed) to the centre of the passband of the MZI, when present, 1550.03 nm.

By continuously controlling the time imbalance of the PIC, not only the signal can be delayed by about a bit-time, but there is also no degradation in transmission quality, nor distortion of the eye-shape (chromatic dispersion), as can be appreciated in Fig. 6.8a–e. Clearly,

Fig. 6.8 Eye diagrams (10 Gbit/s OOK, placed at the center of the spectral response of the PIC, 1550.03 nm) for different values of (continuously) introduced delay, 0 ps (**a**), 22.5 ps (**b**), 50 ps (**c**), 67.5 ps (**d**), 90 ps (**e**). The time domain signal is correctly delayed, while the transmission quality is not impaired. Adapted with permission from [15] ©optica publishing group

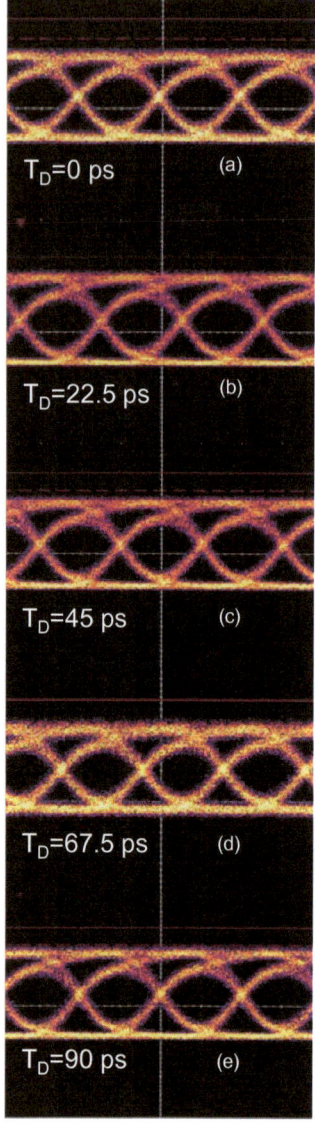

the signal accumulates delay by physically traveling through waveguides, and it is subjected to losses. The differential loss between the shortest and longest path is approximately 3 dB (1.8 dB come from 10% taps, while the remaining amount is due to the combination of propagation losses and TCs' losses). Its evolution (in orange) against the condition of the coupling ratios is reported in Fig. 6.9 (as well as the differential group delay, in blue).

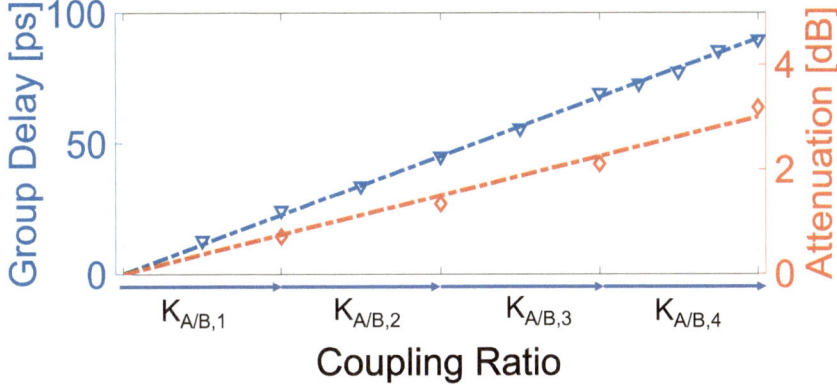

Fig. 6.9 Evolution of the insertion loss (orange markers) and of the group delay (blue markers) introduced by the described delay line, against the coupling ratio of the TCs pairs. The dashed lines represent their theoretical behaviour. Reprinted with permission from [15] ©optica publishing group

6.4.4 Future Development

Spectral and time domain measurements reported in the previous paragraph show that the proposed topology allows for the breakdown of the aforementioned delay-bandwidth product. Fixing a certain m, the $B_{3\,\text{dB}}$ (when it can be defined) remains constant, while the delay increases with M.

$B_{3\,\text{dB}}\tau$ behaviour against m is reported in Fig. 6.10. The solid lines are interpolated measurements results, while the dashed lines come from numerical simulations.

Hence, the exploitation of several nested stages seems to be promising to achieve long delays, while maintaining quite a large bandwidth. However, as already introduced, the differential losses may become a problem.

This may be partially mitigated by employing taps with a lower coupling ratio (for example 1% in place of 10%), at the expense of losing accuracy in PIC tuning. According to Eq. (6.3), a miscalibration of m is linearly proportional to an error in the introduced delay. Another suitable solution could be the use of transparent sensors (without increasing the complexity of read-out electronics), such as those proposed in [17, 18].

Another interesting development for this kind of PIC would be the widening of the operational wavelength range. As already demonstrated in the previous paragraph, the device can operate on a span larger than 60 nm (only limited by the available instrumentation). If there is the necessity to increase this range, to include for example, the S-band and the L-band, the device may not work properly. In these spectral regions, very far from 1550 nm, the TCs might not reach the cross condition, due to the directional couplers dependence on wavelength. This means that portions of the optical signals travel through waveguides that, for the correct operation of the PIC, should be isolated. In fact, the spurious signals traveling through shorter paths have a detrimental effect on transmission quality, causing Intersymbol

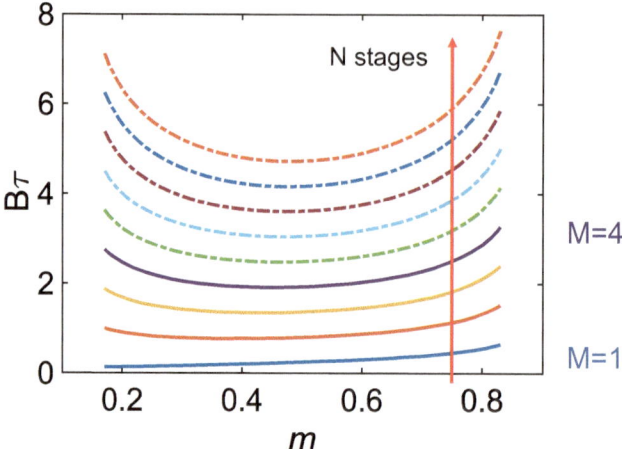

Fig. 6.10 Measured bandwidth-delay product against m for the considered topology, with different number of stages, from $M = 1$, to $M = 4$. For $m > 0.83$ and $m < 0.17$ a 3 dB-bandwidth cannot be defined. The benefit of the architecture is evident. The dashed lines are simulated expected behaviours for $M > 4$, specifically from $M = 5$ to $M = 9$. Reprinted with permission from [15] ©optica publishing group

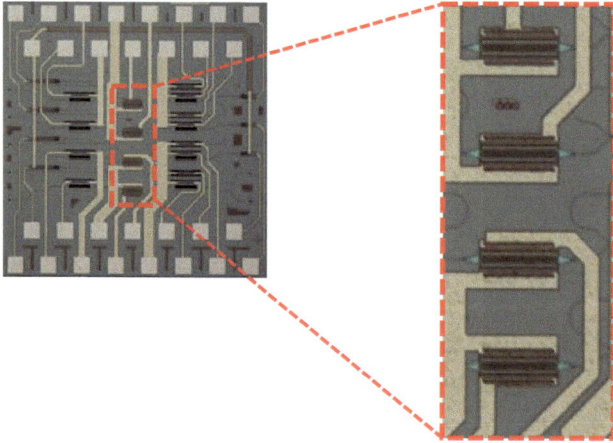

Fig. 6.11 Top-view microphotograph of the p-i-n junctions implementing VOAs surrounding the central waveguides of the PIC

Interference (ISI). This unwanted situation could be effectively avoided by turning on the VOAs (which are not needed for the wavelength range considered in the measurements) and attenuating spurious signals propagating through the central waveguides, but increasing the total losses of the entire PIC. Detailed microphotograph of those VOAs can be found in Fig. 6.11.

6.5 Two-Stage Delay Line

The second delay line topology reported in this book has been theoretically conceived [19], but has never practically realized, characterized and tested.

6.5.1 Topology

This architecture is sketched in Fig. 6.12a. It consists of two cascaded MZIs. Again key building blocks are TCs (controlled by the heaters highlighted in red, for further details see Sect. 6.3), conveniently labeled as K_1, K_2 and K_3.

The MZIs' imbalance is 7.7 mm, so that the single interferometer has a FSR of 10 GHz (and a time imbalance of 100 ps).

The heaters (H_i, with $i = 1, 2$, highlighted in orange) placed above the shorter arm of the MZI allow precise central frequency matching.

The optical Input-Outputs are implemented by means of grating couplers. Immediately after each TC, there is a 1% tap coupler that terminates with a grating coupler.[7] Reading the optical power coming out from those sub-structures (labeled P_1, P_2, P_3 in agreement with TCs' labels) is useful to properly calibrate the PIC.

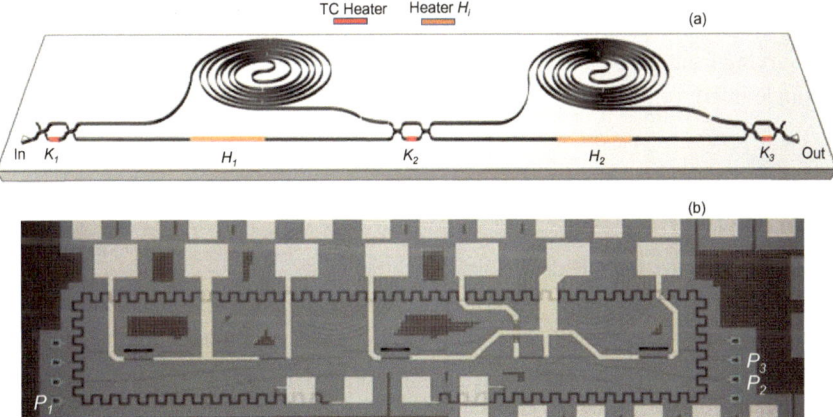

Fig. 6.12 **a** Schematic of the 2-stages delay line. K_1, K_2 and K_3 are TCs. The heaters controlling TCs are highlighted in red and those above the imbalances are highlighted in orange (H_i, with $i = 1, 2$). The in-out grating couplers are also shown. **b** Top-view microphotograph of the implemented PIC

---

[7] Ge-PDs could be and should be put in place of grating couplers to ensure a more compact footprint and a simpler setup (i.e., with the smallest number of optical Input-Output possible).

The device has been realized in the very same platform of the PIC described in Sect. 6.4 (including the waveguide cross section). The top-view microphotograph is shown in Fig. 6.12b. The footprint is 0.5 mm by 1.9 mm.

6.5.2 Device Operation

Even in this case, the considered circuit effectively overcomes the delay-bandwidth constraint. For the previous delay line, a fine coupling ratio control enables a fine and continuous group delay tuning.

Specifically, to effectively exploit the functionalities of this kind of PIC, two conditions must be fulfilled. First, the central frequency of the two MZIs must be the same. Furthermore, the following relations must hold (for power coupling ratios of TCs)

$$K_1 = K_3 = K = \sin^2(\theta), \tag{6.6}$$

$$K_2 = \sin^2(2\theta), \tag{6.7}$$

where θ is a proper control signal for the TCs (i.e., in our case the imposed phase shift).

Under these assumptions, it has been demonstrated that the total group delay is given by [19]

$$\tau = \tau_0 + (2)(K)(T_{MZ}), \tag{6.8}$$

with K varying from 0 to 1 (coherent with definitions given in this Chapter, T_{MZ} is referred to the single interferometer).

6.5.3 Testing Procedure

The testing procedure, for this device, is straightforward.

As already done for the other topologies presented in this book, the electrical validation must first be performed by using the same procedure and equipment widely discussed.

In this case, there are no integrated PDs or p-i-n junctions, thus I–V curves (Fig. 6.13a) are only related to resistive heaters. The thermal cross-talk can be considered negligible (the heaters are ≈ 0.5 mm from each other), but there is still the need for serial testing of TCs.

While performing current-voltage measurements and reading the optical power at P_1, P_2 and P_3, the voltage-coupling ratio characteristics are obtained (Fig. 6.13b).

Optical tuning, instead, is executed following these steps[8]:

---

[8] During the execution of the recipe, the device temperature is stabilized at 25°C, by means of a TEC.

Fig. 6.13 a Current-voltage characteristics of the heaters embedded in the PIC. **b** Voltage-optical Power characteristics of the TCs, constituting the described delay line

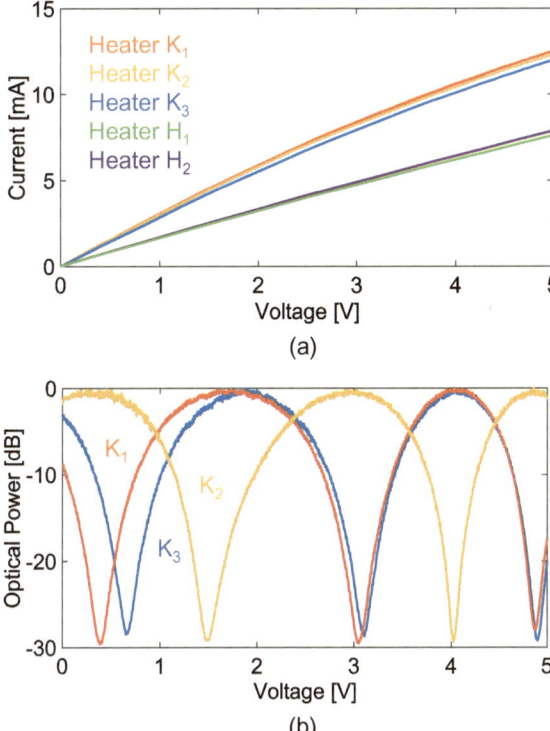

1. Once the input signal (monochromatic source) is coupled to the DUT In port, and the desired delay value (T_D the additional delay with respect to that of the reference path) is stated, K_1 K_2 and K_3 are calculated as:

$$K_1 = K_3 = K = \frac{T_D}{2T_{MZ}}, \tag{6.9}$$

$$K_2 = \sin^2(2\arcsin(\sqrt{K})). \tag{6.10}$$

2. Using a gradient descent approach, the power sensed after Input and central TCs is minimized, so that the optical wave travels through the shortest path, until reaching K_3. At this stage P_1 and P_2 should be almost the same (net of TCs losses, taps mismatch and propagation losses).

3. By means of a simple discrete PI controller, K_3 reaches its setpoint (the bandwidth of this controller is limited only by the thermo-optic time constant). P_3, read at TC output, should satisfy the following relation, $K_3 = P_3/P_2$. By using this ratio as a controlled variable, tuning eliminates input power fluctuations.

4. Step 3 is repeated for K_2 (using the ratio P_2/P_1 as the controlled variable) and K_1 (using P_1 as the controlled variable).

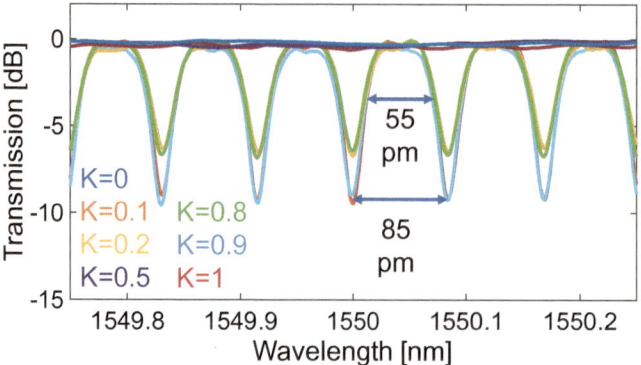

Fig. 6.14 Measured amplitude spectra for the considered delay line with K = 0, 0.1, 0.2, 0.5, 0.8, 0.9, 1, around 1550 nm. FSR is 85 pm, while 3 dB-bandwidth is 55 pm, approximately 25% more than a single MZI

5. The overall output power is maximized, by turning on the two phase shifters placed above the shorter arms of the MZIs.

The entire test lasts approximately 600 ms (500 ms for the electrical part and less than 100 ms for the optical part) and no spectral estimations are needed.

This "tuning for testing" technique accuracy depends on actuators accuracy (0.5 mV with our hardware) and on the optical power read-out reliability.[9] By using our equipment, the time-domain-precision of group delay is ≈ 0.1 ps. If the device is compliant with specifications, the working points of the actuators can be stored in a LUT.

The effectiveness of the algorithm has been validated by means of both spectral and time domain measurements, exploiting the setups described in Sect. 6.5.3.

Spectral responses for $K = 0.1, 0.2, 0.5, 0.8, 0.9, 1$ are reported in Fig. 6.14. 3 dB-bandwidth (55 pm or 6.8 GHz) is $\approx 25\%$ larger than the bandwidth of the a single MZI with the same geometrical imbalance of the single stage of this delay line. The notches are very shallow (i.e., <5 dB in depth) for K spanning from 0.2 to 0.8.

The testing technique (for the same values of K) has also been executed at wavelengths far from 1550 nm, specifically around 1520 nm and 1575 nm, as reported in Fig. 6.15a and b, respectively. The overall FSR passes from 80 pm (10 GHz, at 1520 nm) to 90 pm (11.5 GHz, at 1575 nm), while bandwidth from 50 pm (6 GHz, at 1520 nm) to 60 pm (7.5 GHz, at 1575 nm).

---

[9] Since in this case we use external PMs, this is not a limiting factor.

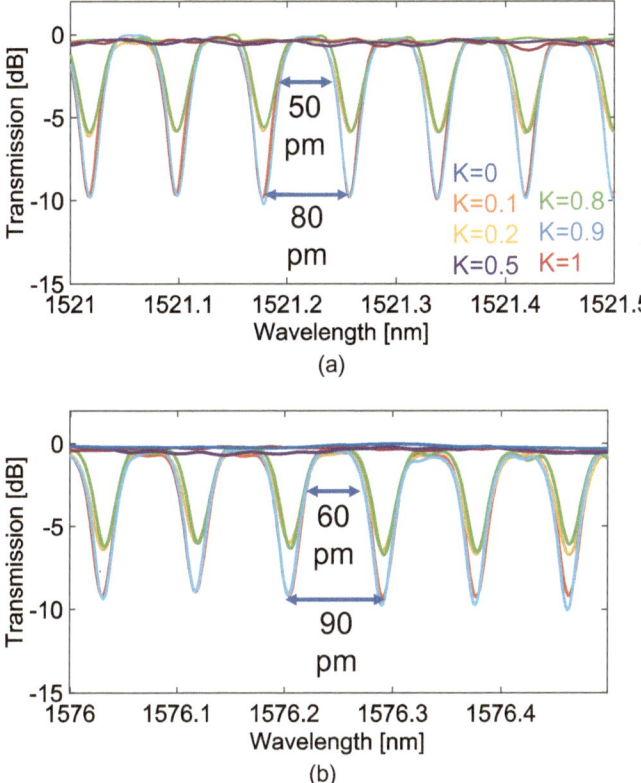

Fig. 6.15 Measured amplitude Spectra for the considered delay line with K=0, 0.1, 0.2, 0.5, 0.8, 0.9, 1, around 1520 nm (**a**) and around 1580 nm (**b**). FSR passes from 80 to 90 pm, while 3 dB-bandwidth passes from 50 to 60 pm

Furthermore, the time imbalance can be continuously tuned and it is two times longer (i.e., 100 ps) than that of a single MZI, as shown in Fig. 6.16a–c. These eye-diagram measurements were performed using the 10 Gbit/s probing signal, $K = 0, 0.5, 1$ (flat spectral response). Even in this case, we can state that the quality of transmission is not impaired. In fact, the differential insertion loss of the whole PIC (between the shortest and the longest paths) is approximately 3.2 dB. Each stage introduces an attenuation of approximately 1.6 dB, mostly due to propagation losses (these waveguides are ≈ 2 dB/cm lossy). Clearly, having fewer lossy structures would enhance the performance of the PIC, enabling even longer arms and delays, at the expense of narrower FSR.

Even for this architecture, the delay-bandwidth product constraint has been overcome.

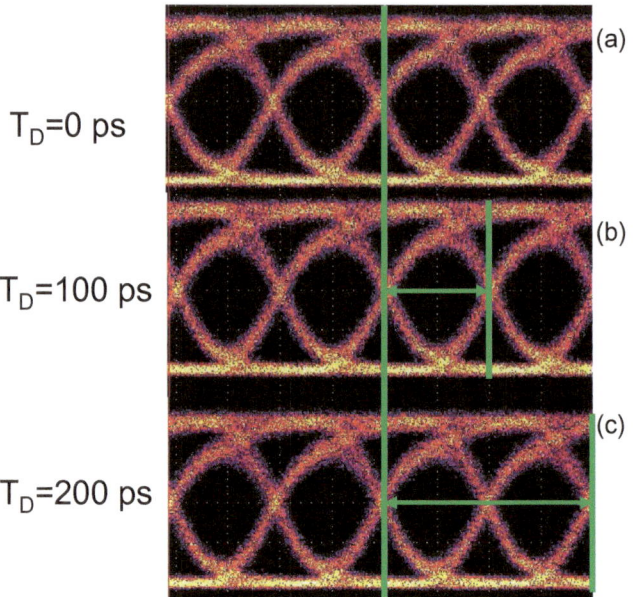

T_D=0 ps

T_D=100 ps

T_D=200 ps

Fig. 6.16 **a** Eye diagrams (10 Gbit/s OOK, placed at 1550.03 nm) for different values of (continuously) introduced delay, 0 ps (**a**), 100 ps (**b**), 200 ps (**c**). The time domain signal is correctly delayed, while the transmission quality is not impaired

References

1. A. Melloni, F. Morichetti, C. Ferrari, and M. Martinelli, "Continuously tunable 1 byte delay in coupled-resonator optical waveguides," *Optics Letters*, vol. 33, no. 20, 2008.
2. D. Spirit, A. Ellis, and P. Barnsley, "Optical time division multiplexing: systems and networks," *IEEE Communications Magazine*, vol. 32, no. 12, 1994.
3. F. Xia, L. Sekaric, and Y. Vlasov, "Ultracompact optical buffers on a silicon chip," *Nature Photonics*, vol. 1, no. 1, 2006.
4. J. F. Diehl, J. M. Singley, C. E. Sunderman, and V. J. Urick, "Microwave photonic delay line signal processing," *Applied Optics*, vol. 54, no. 31, 2015.
5. C. Tsokos, E. Andrianopoulos, A. Raptakis, N. Lyras, L. Gounaridis, P. Groumas, R. B. Timens, I. Visscher, R. Grootjans, L. S. Wefers, D. Geskus, E. Klein, H. Avramopoulos, R. Heideman, C. Kouloumentas, and C. Roeloffzen, "True Time Delay Optical Beamforming Network Based on Hybrid Inp-Silicon Nitride Integration," *Journal of Lightwave Technology*, vol. 39, no. 18, 2021.
6. R. Rotman, M. Tur, and L. Yaron, "True Time Delay in Phased Arrays," *Proceedings of the IEEE*, vol. 104, no. 3, 2016.
7. C. Caucheteur, A. Mussot, S. Bette, A. Kudlinski, M. Douay, E. Louvergneaux, P. Mégret, M. Taki, and M. Gonzalez-Herraez, "All-fiber tunable optical delay line," *Optics Express*, vol. 18, no. 3, 2010.

8. M. Yessenov, B. Bhaduri, P. J. Delfyett, and A. F. Abouraddy, "Free-space optical delay line using space-time wave packets," *Nature Communications*, vol. 11, no. 1, 2020.

9. G. Lenz, B. Eggleton, C. Madsen, and R. Slusher, "Optical delay lines based on optical filters," *IEEE Journal of Quantum Electronics*, vol. 37, no. 4, 2001.

10. A. Melloni, A. Canciamilla, C. Ferrari, F. Morichetti, L. O'Faolain, T. Krauss, R. D. L. Rue, A. Samarelli, and M. Sorel, "Tunable Delay Lines in Silicon Photonics: Coupled Resonators and Photonic Crystals, a Comparison," *IEEE Photonics Journal*, vol. 2, no. 2, 2010.

11. J. Cardenas, M. A. Foster, N. Sherwood-Droz, C. B. Poitras, H. L. R. Lira, B. Zhang, A. L. Gaeta, J. B. Khurgin, P. Morton, and M. Lipson, "Wide-bandwidth continuously tunable optical delay line using silicon microring resonators," *Optics Express*, vol. 18, no. 25, 2010.

12. M. Petrini, M. Milanizadeh, F. Morichetti, and A. Melloni, "Dynamic mitigation of nonlinear effects in a silicon photonic add-drop filter," *Optics Letters*, vol. 46, no. 19, 2021.

13. P. Zheng, X. Xu, D. Lin, P. Liu, G. Hu, B. Yun, and Y. Cui, "A wideband optical beam-forming chip based on switchable optical delay lines for Ka-band phased array," *Optics Communications*, vol. 488, 2021.

14. D. Melati, A. Waqas, Z. Mushtaq, and A. Melloni, "Wideband Integrated Optical Delay Line Based on a Continuously Tunable Mach–Zehnder Interferometer," *IEEE Journal of Selected Topics in Quantum Electronics*, vol. 24, no. 1, 2018.

15. M. Petrini, S. Seyedinnavadeh, V. Grimaldi, M. Milanizadeh, F. Zanetto, G. Ferrari, F. Morichetti, and A. Melloni, "Variable Optical True Time Delay Breaking Bandwidth-Delay Constraints," *Optics Letters*, vol. 48, no. 2, 2023.

16. "PDK of AMF, Advanced Micro Foundry, Singapore, [online]. available: www.advmf.com."

17. S. Grillanda, M. Carminati, F. Morichetti, P. Ciccarella, A. Annoni, G. Ferrari, M. Strain, M. Sorel, M. Sampietro, and A. Melloni, "Non-invasive monitoring and control in silicon photonics using CMOS integrated electronics," *Optica*, vol. 1, no. 3, 2014.

18. A. Perino, F. Zanetto, M. Petrini, F. Toso, F. Morichetti, A. Melloni, G. Ferrari, and M. Sampietro, "High-sensitivity transparent photoconductors in voltage-controlled silicon waveguides," *Optics Letters*, vol. 47, no. 6, 2022.

19. A. Waqas, D. Melati, and A. Melloni, "Cascaded Mach–Zehnder Architectures for Photonic Integrated Delay Lines," *IEEE Photonics Technology Letters*, vol. 30, no. 21, 2018.

Conclusion

In this work, techniques and recipes to perform effective Photonic Testing were discussed.

First, the design, implementation and characterization of novel PICs (e.g., TOADM and true time delay lines) were presented.

Then, since we are dealing with programmable and reconfigurable architectures, embedding several actuators and sensors, we state the importance of electrical testing, which constitutes a pre-condition for the validation of the whole device. To do so, we propose a suitable hardware infrastructure, consisting of a commercial controller, a custom analog stage, for signal conditioning, and a PC for electrical excitation of the PIC under test. By using this equipment, a fast method to estimate the thermal coupling among the actuators was demonstrated.

Due to the complexity of these architectures and the impact of fabrication imperfections on the constituting building blocks, we state that a fast "pre-calibration" is mandatory before performing any kind of optical testing (for example in the time and/or frequency domain [1]). We also validate the techniques and methods for this tuning (executed exploiting the electronics mentioned above). If the two (optical and electrical) souls of the device are compliant with specifications, the state of the actuators of the DUT can be stored in the LUT, which can be updated in case of variations in the environment (e.g. temperature drifts) or different user needs. In the case of photonic filters, we show how to update the LUT in the presence of high-intensity signals triggering nonlinear phenomena.

The very aim of this book is the introduction of a framework and the drawing of guidelines for the validation of novel (programmable) photonic circuits. The proposed solutions are intended to improve the throughput of testing (in the sub second range for an entire PIC) and are effective in terms of cost per number of (electrical and optical) channels.

However, the path to bring PIC testing at the same level of maturity as EIC testing is still quite long (this consideration holds for many other aspects of photonic supply chain, even though much effort has been made in these years). We believe that many gaps could be filled

© The Editor(s) (if applicable) and The Author(s), under exclusive license to Springer 133
Nature Switzerland AG 2025
M. Petrini, *Mixed-Signal Generic Testing in Photonic Integration*, Synthesis Lectures
on Digital Circuits & Systems, https://doi.org/10.1007/978-3-031-60811-7

with the joint contribution of all the communities and players involved in Photonics, from the foundries to the final users.

For example, the lack of standardization (especially, but not only, in Testing) is now limiting the development of photonics and it is dramatically increasing the total cost of a PIC. There is an urgent need for well defined standards, regarding equipment specifications, automation infrastructure (electro-optical probing systems and their movements), acceptance thresholds and software application programming interfaces (APIs).

There should also be, guidelines to design photonic chips, making the testing activities easier. This approach goes under the name of DFT and in the recent decades has been one of the key to the success of microelectronics, reducing the number of issues and the cost per tested EIC.

To make a practical example, the layout of a PIC, regardless of the implemented functionalities, should be shared among the all the players in photonics. In this way, by fixing the position of the electrical pads and the optical I/Os, usage (and design) of many different PCs can be avoided.

Concerning, instead, the automation and the mechanical alignments, the standards already present in the microelectronic industry (in terms of precision, accuracy, repeatability, velocity and applied forces) should be inherited and guidelines for optical coupling (for all possible combinations, horizontal/vertical and passive/active) should be tracked.

The electronics that perform these tests should also be standardized. The analog frontends should be able to read/provide proper signals to/from the PIC within pre-defined ranges (voltages/currents). The specifications about the digital controllers and the central units should be clear in terms of performance, communication protocols, nature of the operative systems (whether real-time or not).

Standard programming patterns and application programming interfaces (APIs) should be defined.

Data should be stored in proper databases, in a standardized format [2].

Finally, the PICs should be somehow categorized (according to their platform, their functionality and so on) and for each sub-set commonly accepted thresholds should be defined (i.e., an MZI mesh and a WDM filter are different, so different parameters must be measured and different pass/fail rules must be followed).

These practices could be and should be immediately pursued. Indeed, there are interesting examples in the literature towards this direction [2]. Nevertheless, we can suggest also longterm actions, that could enhance photonic testing in the future.

First, thanks to increasing standardization, we should foresee a quite important transition, that will change the paradigm of testing: from "platform oriented" (such as the contents of this book) to "platform agnostic" testing. These new trends will enable the shift from a "white box" approach to a "black box" approach. This will open new possibilities, such as the use of Machine Learning and Artificial Intelligence, for classification and validation purposes [3], which will be completely general and not related to a specific architecture and/or technology.

Another point, that must be addressed, is the development of techniques for self-diagnostics of the PIC. This is fundamental for the transition from a "photonic device" to a "system-on-chip". By developing this approach, failures can be monitored and reported in real time, not only during the different stages of production, but also through the normal operation of the device, leading to the so called "life-long testing".

We believe that these milestones (i.e., standardization, DFT, platform agnostic testing and self-diagnostics) once achieved, would boost activities related to the validation of the integrated devices, and would also be beneficial for the whole Photonic eco-system.

Future Directions

Concerning the content of this book, the following aspects are under further investigation.

- A polarization insensitive-four channels-TOADM is now becoming the new DUT. The techniques and recipes discussed in this work are being applied to this challenging PIC. Since this new photonic circuit will be implemented in the same platform as the previous ones, it could also be used to deeply investigate the nonlinear behaviour of the exploited waveguides and materials.
- In Chap. 3 we discussed a proper electronic infrastructure, aimed at low-frequency testing of silicon PICs. However, higher frequency testing (i.e., in the range of GHz or tens of GHz) cannot be neglected. To accomplish this task, not only the analog signal processing (and PXI board set) should be layouted from scratch, but also the probe card should be completely re-designed, avoiding electrical crosstalk among needles. In this scenario, (electrical) testing signals would become much more sensitive to mechanical contact.
- A novel version of the PC is under preparation. This will be an hybrid version, to perform, at the same time, electrical and optical access of the DUT, in a wafer-level-testing scenario. To do so, the mechanical alignment system is being partially modified. Different approaches have been considered so far, exploiting the trenches (that are usually etched) between consecutive PICs. The alignment routine could be active (by means of accurate piezo-actuators integrated in the PC itself) or passive.
- Methods and techniques for optical testing are introduced in Chap. 4. One of the most important achieved results is that of "filter cloning". We show how to replicate, with the DUT, the amplitude transfer function of a specific REF, under certain conditions. The technique could be expanded, including the "phase replication" of the REF. This would be quite important for a large set of devices, such as all-pass filters.
  Moreover, the approach is being validated on multiport devices.
- Concerning optical true time delay lines (and in particular that constituted by four nested MZIs), the next steps have been discussed in Chap. 6. A wider operational wavelength range will be demonstrated, also recurring to the use of embedded VOAs. The design for a new version, with a higher number of stages and equipped with transparent detectors (to reduce losses as much as possible), is under preparation.

Time Domain Multiplexing

There could be some situations in which the parallel electrical testing (i.e., the parallel acquisition of several I-V curves), with the hardware described in Chap. 3, is too slow in terms of throughput (mainly due to the analog interface board between the PXI platform and PIC). As we have seen in the main text of this dissertation, there should be a trade off between the electrical bandwidth of exploited EICs, the number of channels and the overall cost of the instrumentation. Thus, due to the topology of the photonic circuit, serial electrical testing is needed, and the overall testing time scales with the number N of embedded devices in the PIC.

Another point to be highlighted is that, following the recipes in Chap. 3, DUT overheating may occur (since it is turned on at a sufficiently high voltage for several ms).

A suitable solution to overcome these issues could be the use of a well known technique TDM. By using a fast single-channel SMU, the single electrical device under test could be driven with a train of short voltage pulses, with increasing amplitude. At the same time, by using the same SMU, the current absorbed by the load could be sensed. By sampling the provided voltages and reading currents at sufficiently high frequency (i.e., sampling time at least ten times shorter than pulse width, as a rule of thumb), the I-V curves could be properly built (provided that the electrical bandwidth of the DUTs is wide enough for the chosen pulses).

Between a pulse and the following one, there is usually a guard time, useful, to let the DUT cool down. During this "void-time" interval, another short-pulse-waveform can be generated by the SMU and routed to another DUT (not affected by spurious coupling effects), by means of a 1-to-N electrical MUX.

© The Editor(s) (if applicable) and The Author(s), under exclusive license to Springer Nature Switzerland AG 2025
M. Petrini, *Mixed-Signal Generic Testing in Photonic Integration*, Synthesis Lectures on Digital Circuits & Systems, https://doi.org/10.1007/978-3-031-60811-7

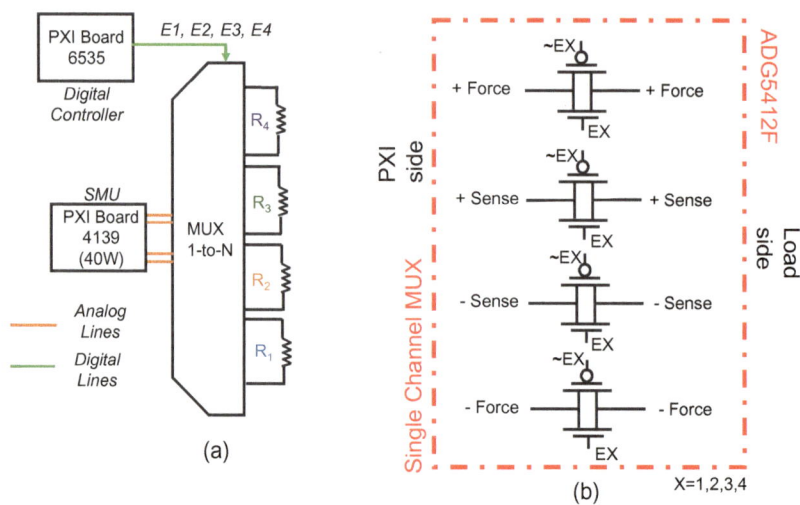

Fig. A.1 **a** Block scheme exploited to set up the time domain multiplexing approach. **b** Circuit scheme of the EIC, used as a building block for the MUX. The enabling signal and its negation are reported

The block scheme is shown in Fig. A.1a. The system has been effectively implemented by using the PXI board 4139 [4] as SMU (capable of generating rectangular pulses of minimum of 10 μs, configured to provide voltages between ±6 V and currents ±20 mA, with a resolution of 1 μV and 10 nA, respectively), and a bank of fast solid state switches (single EIC is $ADG5412F$, from Analog Devices [5]) as MUX, configured as in the circuit of Fig. A.1b and actually arranged in a custom PCB (four EIC in total). The transition time of these FET-switches is ≈ 400 ns (and the passband is > 10 MHz), and in each EIC package there are four independent pass-transistors, which are dedicated to a single DUT. The SMU 4139, in fact, offers the possibility of performing 4-wires measurement (also known as *Kelvin Sensing*), which are more accurate than those performed with 2 wires. The sensing wires are placed as close as possible to the PC needles, contacting the DUT.

The switches state is controlled by means of a CMOS digital signal, 0 − 3.3 V, provided by the PXI board 6535 [6] (and hardware-synchronized with 4139).

The time domain behaviour of the testing pulses and of the digital signals enabling a specific DUT measurement is reported in Fig. A.2a (in the inset, the detailed pulse shape is shown). We use these waveforms to test the four heaters embedded in a generic PIC (not thermally coupled). The I-V curves of those DUTs are shown in Fig. A.2b. In this case we choose a pulse width of 10 μs, a duty cycle of 1% (the void time is 990 μs), and a voltage span between 0 and 5 V, with $\Delta V = 0.1$ V (50 datapoints). After 250 μs from the end of the i-th pulse [with an amplitude $V_i = (i)(\Delta V)$, where i is the ordinal number, counting the number of pulses per heater] exploited to test the first heater, and another identical pulse

Fig. A.2 a Time domain behaviour of the analog and digital signals. The colours of the digital waveforms are matched to the different DUTs, while the signal provided by the SMU is in black. The duty cycle is 1% ($T_{ON} = 10\,\mu s$ and $T_{OFF} \approx 1$ ms). **b** I-V curves of the four DUTs

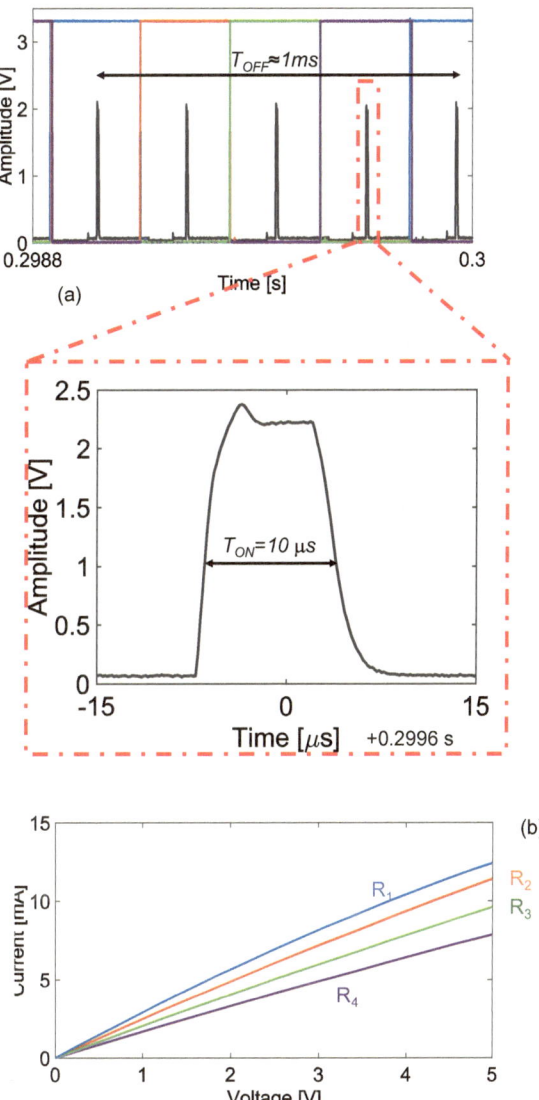

excites the second heater. After $250\,\mu s$ from this event, the same occurs for the third and finally for the fourth heater. And so on, for the (i+1)-th pulse.

Between two consecutive pulses, the SMU is selectively connected to a specific heater, while the others are grounded.

In this way, in 50 ms four heaters are completely tested, in terms of I-V characteristics. Hence, throughput is enhanced by one order of magnitude with respect to the results shown in Chap. 3, and there is even room for improvement. By increasing the number of switches, in fact, a larger number of DUTs can be tested in the same integral time.

Appendix

B

Interferometer Based Filters

Given the extensive treatment of photonic filters in the main text, this appendix contains formal definitions of the filter terminology (adopted in [7]) and a general overview of MRR-based devices (with particular focus on Vernier topologies) and MZI-based devices. The section concludes with a complete theoretical treatment of the concepts behind the filters discussed in Chaps. 2 and 6, respectively.

General Terminology

A generic transfer function of a bandpass filter, operating in the WDM range, is shown in Fig. B.1. According to the curves depicted in the picture, the following features can be defined:

- Insertion Loss (IL): amount of attenuation introduced by the device, at a given frequency.
- Bandwidth (B): range of frequencies, where the attenuation is less (in absolute value) than a certain threshold (usually 3 dB, thus the bandwidth is labeled as $B_{3\,dB}$). This figure of merit is useful for band-pass transfer functions (such as In-Drop in an MRR filter).
- Return Loss (RL): the minimum insertion loss expected at a specific range of frequencies. This figure of merit is particularly useful for band-stop transfer functions (such as In-Through in an MRR filter, to evaluate rejection at resonance).
- Isolation at a specific frequency (I): insertion loss of the passband at a certain frequency far away from its own center by an amount Δf (usually, in the main text, $\Delta f = 50$ GHz, thus isolation is labeled as $I_{50\,GHz}$).

© The Editor(s) (if applicable) and The Author(s), under exclusive license to Springer Nature Switzerland AG 2025
M. Petrini, *Mixed-Signal Generic Testing in Photonic Integration*, Synthesis Lectures on Digital Circuits & Systems, https://doi.org/10.1007/978-3-031-60811-7

Fig. B.1 Spectral response of a generic filter, for both passband (in red) and stopband (in blue) ports

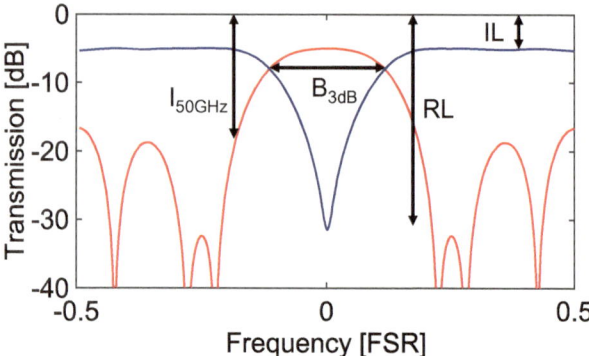

- Operational Wavelength Range: range of frequencies, in which device performances (according to the figures of merit listed above) can be considered acceptable.

These definitions are intended to better clarify the text and constitute a nomenclature for the set of specifications discussed in the last subsection.

Interferometer Based Filters

Ring Resonator Add/Drop Filter

The simplest Optical Add/Drop filter can be built exploiting an MRR. This is an example of an IIR device, with a single pole response (In-Drop). A detailed sketch of this circuit (constituted by two bus-waveguides coupled by means of a circular one) is reported in Fig. B.2a. Its (complex) transfer functions, at Through and Drop ports, are respectively:

$$H_{Through} = \frac{r_1 - r_2\alpha e^{-j\beta L_r}}{1 - \alpha r_1 r_2 e^{-j\beta L_r}}, \tag{B.1}$$

$$H_{Drop} = -\frac{t_1 t_2 \sqrt{\alpha} e^{-\frac{j\beta L_r}{2}}}{1 - \alpha r_1 r_2 e^{-j\beta L_r}}, \tag{B.2}$$

where L_r is the circumference length, $\beta = \frac{2\pi n_{eff}}{\lambda}$ is the propagation constant (where n_{eff} the effective refractive index), α is the round trip loss, t_1 and t_2 are the cross coupling coefficients, while r_1 and r_2 are the self coupling coefficients. Thus, $[[t_1^2, r_1^2]$ and $[t_2^2, r_2^2]$ are the power splitting ratios of the two couplers. In lossless conditions the following relations hold: $t_1^2 + r_1^2 = 1$ and $t_2^2 + r_2^2 = 1$.

The round-trip losses and the coupling ratios univocally define the position of the pole. The closer the pole is to the unit circle, the narrower the passband and the higher the RL at

Fig. B.2 a Sketch of a single MRR. The field coupling coefficients are highlighted. **b** In-Through spectral responses (dashed lines) and In-Drop spectral responses (solid lines), for different values of the coupling ratio

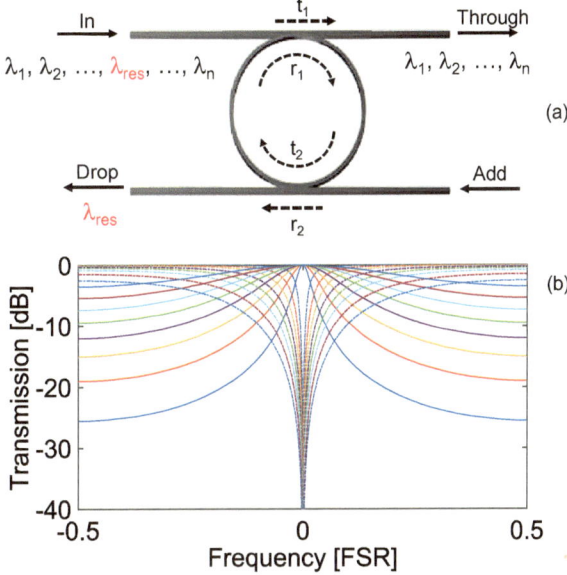

resonant frequencies. The condition, according to which round-trip losses match coupling (i.e. $r_2\alpha = r_1$), is called *critical coupling*.

Some parameters define the features of this kind of filter:

- Resonance frequency ($\lambda_{res,N}$): wavelengths that show constructive interference, after travelling for one roundtrip, at the output port of the filter (Drop). The resonance condition is satisfied when:

$$\beta L_r = \frac{2\pi}{\lambda} n_{eff} L_r = 2N\pi,\qquad (B.3)$$

which can be solved as:

$$\lambda_{N,res} = \frac{n_{eff} L_r}{N},\qquad (B.4)$$

where N is the resonance order ($N = 1$ is the *fundamental order*).

- Free Spectral Range (FSR): distance, in wavelength (or frequency) domain, between two consecutive resonances. The formal definition is:

$$FSR(\lambda) = \lambda_N - \lambda_{N-1},\qquad (B.5)$$

which can be equivalently expressed as:

$$FSR(f) = \frac{c}{n_g L_r},\qquad (B.6)$$

where n_g is the group index and c is the vacuum speed of light.

- Bandwidth: in the case of lossless and identical power coupling ratios (K, between the ring and waveguides), it can be defined as

$$B = \frac{FSR}{\pi} \frac{K}{\sqrt{1-K}}.$$ (B.7)

- Finesse: defines the frequency selectivity of the filter. Analytically, it is:

$$Finesse = \frac{FSR}{B}.$$ (B.8)

- Maximum group delay (τ_p), which represents the time spent by the optical signal inside the filter (in the resonance and assuming $r_1 = r_2 = r$):

$$\tau_p = T\left(\frac{1+r}{1-r}\right)$$ (B.9)

being T the round-trip time of the ring.

To select only one channel in the WDM spectral range, the FSR should be larger than this wavelength interval. According to Eq. (B.6), the FSR is proportional to L_r^{-1}, this would require a very small radius, which may lead to an extremely high bending losses (in Silicon Photonics the minimum radius should be $>7\,\mu m$).

At the same time other requirements should be matched, such as the filter bandwidth, which should be larger than the single WDM channel bandwidth, to avoid a signal distortion or the killing of some harmonics. The bandwidth can be designed by acting on the coupling ratios. However, the bandwidth optimization worsens the RL, so a trade-off should be found. The transfer functions (Through and Drop ports), for different values of K (supposed to be equal for the two bus waveguides) of a single Add/Drop MRR are shown in Fig. B.2b.

Filters with multiple poles can be obtained by cascading several MRRs. In so doing, many degrees of freedom can be exploited to match the specifications in terms of spectral parameters, at the expense of a more complex design and calibration.

Vernier Configuration

Coupled MRR architectures offer appealing features for filtering applications. However, their FSR is limited by the constraints stated in the previous paragraph, if all the rings have the same circumference length. This issue could be overcome by combining rings with different radii (Vernier Configuration).

Let us consider a second order filter, consisting of two rings, with different radii (but the same n_g), R_1 and R_2. Such a device is shown in Fig. B.3. The total FSR of this filter is

$$FSR_{tot} = m_1 FSR_1 = m_2 FSR_2,$$ (B.10)

Fig. B.3 Sketch of a second order MRR Vernier filter. The field coupling coefficients are highlighted

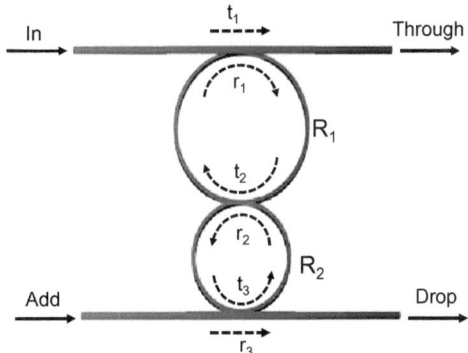

where m_1 and m_2 are *Vernier Coefficients* and are defined as the ratio between the FSR of each single ring and the minimum of the set.

From Eq. (B.10), the following also holds

$$\frac{m_2}{m_1} = \frac{L_2}{L_1}. \tag{B.11}$$

Clearly, to have the largest FSR, m_1 and m_2 have to be coprime. The shortest radius of the series is usually determined by acceptable bending losses.

The simultaneous resonance of all the MRR is given by:

$$\lambda_{res} = \frac{n_{eff} L_{r,1}}{N_1} = \frac{n_{eff} L_{r,2}}{N_2}, \tag{B.12}$$

where $L_{r,1}$ and $L_{r,1}$ are the geometrical circumference lengths, while N_1 and N_2 are the resonance orders.

In presence of a (very) high order filter (composed of several MRRs), a number of strategies to maximize FSR extension have been proposed in the literature.

Mach-Zehnder Interferometer Based Filters

The Mach-Zehnder Interferometer can be also used as building block for complex filters. Its transfer function is sinusoidal, thus it turns to be less selective than the MRR. This is an example of an FIR device, with a single zero. A detailed sketch of this circuit (constituted by two couplers connected by means of two waveguides, having, in general, different length) is reported in Fig. B.4.

Its transfer functions, at the *bar* and *cross* ports, can be computed in the complex domain (considering the situation with symmetric couplers, whose power coupling ratio is K):

$$H = T_c T_L T_c \tag{B.13}$$

being T_c the matrix representation of a generic directional coupler and T_L the phase difference matrix:

$$T_c = \begin{bmatrix} cos(\kappa) & -jsin(\kappa) \\ -jsin(\kappa) & cos(\kappa) \end{bmatrix} \tag{B.14}$$

$$T_L = \begin{bmatrix} 1 & 0 \\ 0 & e^{-j\Delta\phi} \end{bmatrix}, \tag{B.15}$$

where $\kappa = arcsin(\sqrt{(K)})$ and $\Delta\phi$ is the phase umbalance between the two optical paths.

Assuming 50:50 couplers and computing the power transfer, the following equations hold:

$$P_{bar} = sin^2\frac{\Delta\phi}{2} \tag{B.16}$$

$$P_{cross} = cos^2\frac{\Delta\phi}{2} \tag{B.17}$$

It turns out that a relevant degree of freedom for these responses is the phase difference of the signals propagating in the two waveguides, and thus the different optical length (ΔL). This mutual relation can be expressed as:

$$\Delta\phi = \frac{2\pi}{\lambda}n_{eff}\Delta L \tag{B.18}$$

which holds for waveguides made of the same material (usually true for integrated circuits) and neglecting any dependency between the propagation constant and the waveguide curvature.

Therefore, Eqs. B.16 and B.17 become:

$$P_{bar} = sin^2\left(\frac{\pi n_{eff}\Delta L}{\lambda}\right) \tag{B.19}$$

$$P_{cross} = cos^2\left(\frac{\pi n_{eff}\Delta L}{\lambda}\right) \tag{B.20}$$

A plot of the power transfer as a function of wavelength is shown in B.5.

ΔL is also crucial to define important spectral features (referred to the 50:50-directional-couplers-MZI), such as:

- Free Spectral Range (FSR), which, in this case, represents the period of the sinusoid (being $\lambda_{Y,max}$ the position, in wavelength domain, where the sinusoid is maximized)

Fig. B.4 Sketch of a single MZI, having symmetric directional couplers

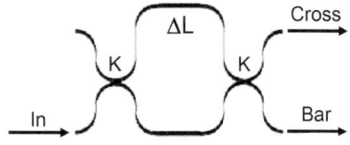

Fig. B.5 Power transfer versus wavelength of an unbalanced MZI, with 50:50 directional couplers

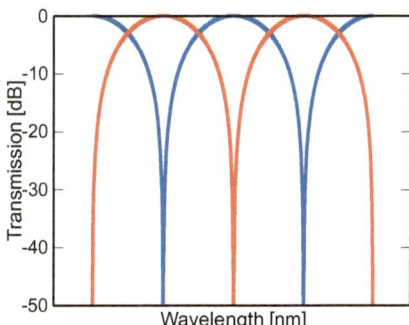

$$FSR(\lambda) = \lambda_{N,\max} - \lambda_{N-1,\max}, \tag{B.21}$$

can be equivalently expressed as:

$$FSR(f) = \frac{c}{n_g \Delta L}, \tag{B.22}$$

where n_g is the group index and c is the vacuum speed of light.

- Bandwidth (B, at −3 dB):

$$B = \frac{FSR}{2} = \frac{c}{2n_g \Delta L}. \tag{B.23}$$

- Maximum group delay (τ_p), which represents the time spent by the optical signal inside the filter (at the peak of the sinusoidal transfer function):

$$\tau_p = FSR^{-1} = \frac{n_g \Delta L}{c} \tag{B.24}$$

This is shorter with respect to the MRR group delay, but that device exhibits a narrower bandwidth.

The spectral response of the MZI can be rigidly shifted (in the frequency or wavelength domain) acting on the difference of the optical path of the two internal waveguides (for example, exploiting thermo-optic effect).

The coupling ratio of the directional couplers (symmetrically controlled) is another degree of freedom to tune the transfer function of the MZI. The farer from the 50:50 situation, the smaller the ER, defined as the ratio between the maximum and the minimum of the transfer function (it tents to infinity, in presence of 50:50 couplers).

Furthermore, the higher the coupling ratio, the longer τ_p. Referring to the sketch in Fig. B.4, when K = 1 the optical signal travels through the longest waveguide and spends more time inside the photonic structure.

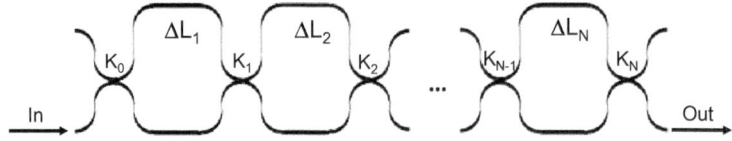

Fig. B.6 Sketch of a lattice MZI structure, having, in general, different directional couplers

Lattice Mach-Zehnder Structures

Cascading a number of MZIs, such as in Fig. B.6, more appealing spectral features can be achieved. In addition, increasing the number of the degrees of freedom, the flexibility of the spectral response increases.

Analytic transfer function can be expressed as:

$$H(\lambda) = \frac{1}{N} \sum_{i=0}^{N-1} c_{Ki} e^{-j\frac{2\pi}{\lambda}i\Delta L_i} \tag{B.25}$$

where c_{Ki} is a quantity that depends on the coupling ratio of the directional couplers, connecting the stages (N, in total).

These structures, if properly synthesized, can act as (amplitude and phase) equalizers, delay lines or WDM filters, having quite large bandwidth and group delay, reduced cross-talk and chromatic dispersion.

References

1. Y. Wang, P. Sun, J. Hulme, M. A. Seyedi, M. Fiorentino, R. G. Beausoleil, and K.-T. Cheng, "Energy Efficiency and Yield Optimization for Optical Interconnects via Transceiver Grouping," Journal of Lightwave Technology, vol. 39, no. 6, 2021.
2. S. Latkowski, D. Pustakhod, M. Chatzimichailidis, W. Yao, and X. J. M. Leijtens, "Open Standards for Automation of Testing of Photonic Integrated Circuits," *IEEE* Journal of Selected Topics in Quantum Electronics, vol. 25, no. 5, 2019.
3. P. Gaur, S. S. Rout, and S. Deb, "Efficient Hardware Verification Using Machine Learning Approach," in 2019 IEEE International Symposium on Smart Electronic Systems (iSES) (Formerly iNiS), IEEE, 2019.
4. PXIe-4139, *Source and Measurements Units*. NI, 2022.
5. ADG5412F, *Fault Protection and Detection, 10 Ohm Ron, Quad SPST Switches*. Analog Devices, 2017.
6. PXIe-6535, *Digital Signal Board*. NI, 2022.
7. A. Melloni and F. Morichetti, *Componenti e Circuiti per le Comunicazioni Ottiche*. 2010.